# RELATIVITY

## An Introduction to the Special Theory

# RELATIVITY

# An Introduction to the Special Theory

**Asghar Qadir**
Department of Mathematics
Quaid-i-Azam University
Pakistan

*Published by*

World Scientific Publishing Co. Pte. Ltd.,
P O Box 128, Farrer Road, Singapore 9128
*USA office:* 687 Hartwell Street, Teaneck, NJ 07666
*UK office:* 73 Lynton Mead, Totteridge, London N20 8DH

**RELATIVITY: AN INTRODUCTION TO THE SPECIAL THEORY**



ISBN 9971-50-612-2

Printed in Singapore by Loi Printing Pte. Ltd.

*Dedicated to*

*my wife*

***RABIYA QADIR***

# PREFACE

This book is based on a series of lectures I gave for M.Sc. students repeatedly over the years at the Department of Mathematics, Quaid-i-Azam University, Islamabad. These lectures evolved according to the needs of the students whom I taught. The background of the students was very different from that of Western students at a similar level. The problem was, keeping the lectures largely self-contained, to develop the background required for Special Relativity more or less from scratch. Of course, this development must be made in a finite time – one semester, to be precise. On the presumption that these problems are faced throughout the Third World, I decided to publish this book with a press that could make it available for the Third World.

It is hoped that this book will be useful, not only in the Third World, but everywhere. It is aimed at an audience making its first real contact with the Special Theory of Relativity. A background of matrices and vectors, of differential and integral calculus and the rudiments of group theory is assumed. Virtually no background in Physics is assumed except for some Classical Mechanics and a nodding acquaintance with the formalisms of Lagrange and Hamilton.

A word of explanation is in order as to why the General Theory of Relativity is not included *at all* in this volume. In Pakistan, this subject is dealt with separately in a one-semester course. The background required for it is more extensive. All things considered, it did not seem useful for most readers to be forced to 'buy' the General Theory when they are only interested in the Special Theory. This is particularly true for students of Physics who intend to work on the experimental side, or even in branches of Theoretical Physics other than Particle Physics, Cosmology, Astrophysics or General Relativity itself. While General Relativity is steadily gaining importance in Physics, it is by no means as basic for it as the Special Theory is.

I have tried to maintain, throughout, a historical perspective of the development of the Theory of Relativity. This is done for two reasons. First, I believe that it helps to build an interest in the subject and give credit where it is due. Second, I believe that it helps to give a better 'feel' for the concepts on which the theory is based. Many nuances are 'lost in non-translation' in carrying forward earlier terms out of their original context.

Finally, I would like to record my indebtedness to my late father, Mr. Manzur Qadir, who introduced me to the pleasures of Relativity; to my Ph.D. supervisor, Professor Roger Penrose, from whom I learnt precision in thinking (particularly in the field of Relativity); to Professor John Archibald Wheeler, who brought home to me the importance of clear and attractive presentation of ideas; and to my numerous students on whom I experimented in an effort to find the best method of teaching Relativity to students with the background available in Pakistan. Thanks are also due to many colleagues and students who refused to 'see the Emperor's new clothes' till they were put on. Of course, my gratitude goes to my family who were neglected because of this work and in particular to my wife, Rabiya Qadir, for giving me continual support and encouragement in writing this book. Finally, I wish to thank Mr. Shabahat Ullah Khan for his excellent typing of the manuscript.

ASGHAR QADIR

# TABLE OF CONTENTS

# Chapter 1

# INTRODUCTION

Relativity theory as it stands may be thought of as the study of motion. In this sense Special Relativity is the theory dealing with uniform motion. It deals with 'kinematics' rather than with 'dynamics' (which deals with the motion of a body experiencing force). Before going on to the subject itself, we will first take a brief look at the history of the theories of motion. In the process we will also need to consider the scientific method as we now think of it and as it used to be thought of.

## 1. Historical Background of Motion

Familiar as we are today with the concepts of acceleration, velocity, time, etc., it is very difficult to understand what motion meant to the ancient mind. At the very start, it could not have been more than the fact that an object occupied different positions at different instants. Thus, a man seen in the village on one occasion and in the forest on another, had moved. Similarly, a lion one ran away from in the forest, seen in the village, had moved. If a stone seen in one part of the village was found elsewhere, it had moved. Those objects that were seen to be able to move of their own volition were called 'animate' objects, while those that could not were called 'inanimate'. If an inanimate object moved, but there was no apparent visible object which had moved it, an invisible personality – a spirit – was supposed to have moved it. Thus, for example, Greek myths abound in 'wood-spirits' or 'dryads' which moved the leaves of trees and 'water nymphs' which caused the motion of water that was seen, but was not due to fish.

As the human mind started to grope for more general causes, which could be impersonal, people came to search for patterns. One of the most important of the ancient formulations was that of Aristotle. He stated some 'self-evident truths', as he saw them, and deduced the observed patterns from them. This fitted in with the view of science as held at that time. We shall look at his 'laws of motion' and the 'self-evident truths' supporting those laws.

It seemed an obvious truth to Aristotle that the most perfect curve, and hence path, is a perfect circle and the most perfect shape is a sphere. Also, that the Heavens are perfect while the Earth is imperfect. Further, that objects tend to return to their place of origin, whether they be animate or inanimate. Now, from the first two principles Aristotle deduced his *law of celestial motion*: 'All Heavenly bodies move in perfect circles, except insofar as they may be made imperfect due to the influence of the Earth, whereby they develop epicycles – the more epicycles the closer they are to the Earth'. Aristotle had stated that the Universe was made up of five 'elements'. The four Earthly elements were, in order of increasing perfection: earth; water; air; fire. The Heavenly element was 'aether'. This belief made more concrete his *law of terrestrial motion*, which states that 'All terrestrial bodies tend to go to their natural state of rest'. This law explains why a stone will fall to the Earth – since it was taken from there and that had been its natural state of rest. Similarly the apple will fall to the ground because the seed from which the apple tree grew had been sowed in the ground. Again on burning wood, when smoke rises and ashes fall; this is because the earthly part is returned to the Earth, while the airy part goes back to the air. In the process, a certain amount of water and fire which were contained in the wood are released. It was, again, self-evident to Aristotle that the more Earthly something was the greater its tendency would be to get back to the Earth and therefore it would fall faster. The Earthly 'elements' are what we would now call solid, liquid, gas, and energy. We will return to the 'aether' later.

The motion of the Heavenly bodies requires further discussion. As regards their motion, there were two types of objects normally in the sky: the fixed stars, which were points of light and moved in perfect circles, and the planets which were larger objects and moved in more eccentric ways. In addition there were comets and meteors which Aristotle identified as atmospheric phenomena, and hence are terrestrial in nature. The Heavenly objects were made of aether, but nearness to the Earth could contaminate them. Thus the more imperfect-seeming Heavenly objects should be closer to the Earth and move more eccentrically. This eccentric motion was given by an epicycle, i.e., a perfect circle whose centre moves in a perfect circle about the Earth. If the degree of contamination increased there would be more epicycles in the orbit of the object. Thus, there would be increasing fiery contamination of those objects which had more epicycles. Now, aether was unchanging and eternal while fire was changing all the time. Thus the Moon, which had the greatest epicycles, changed the most, but still it changed cyclically. The Sun, still with a lot of fiery contamination, but much less changeable (changing with a cycle of a year instead of a month), had fewer epicycles in its orbit. The other planets, again, had very little changeability and few epicycles.

Contemporaneous with Aristotle were scientists who saw the Heavens very differently. They believed that the Moon went round the Earth, but that the Earth and other planets went round the Sun. The major proponent of this view was Aristarchus of Samos. A follower of this view, Eratosthenes of Cyrene, made some beautifully simple observations to deduce the size of the Earth. He noticed that on the shortest day of the year, the shortest shadow cast by an upright stick decreased to vanishing. However, 500 miles due North of Alexandria (where the shadow vanished), a shadow was cast corresponding to an angle of 7° (see Fig. 1). Thus, when the Sun was directly overhead on the equator it was 7° lower at a distance 500 miles due North. Now the ratio of the circumference of the Earth to

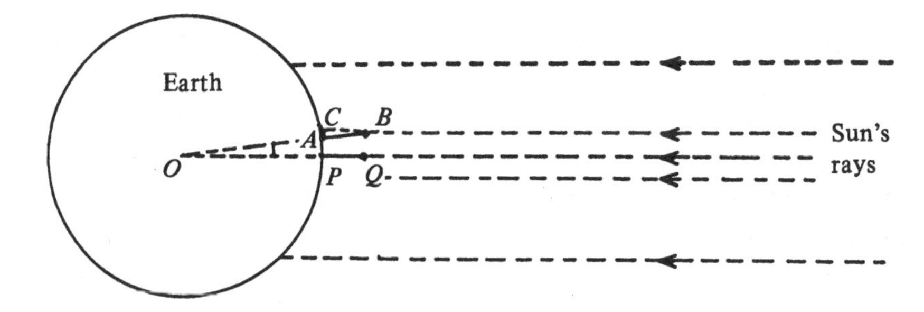


Fig. 1. Eratosthenese's measurement of the Earth's circumference and diameter. The stick $PQ$ casts no shadow. The stick $AB$ casts the shadow $AC$, which gives the angle subtended by $PA$ as 7°. The ratio of the circumference to $PA$ is equal to the ratio of 360° to 7°.

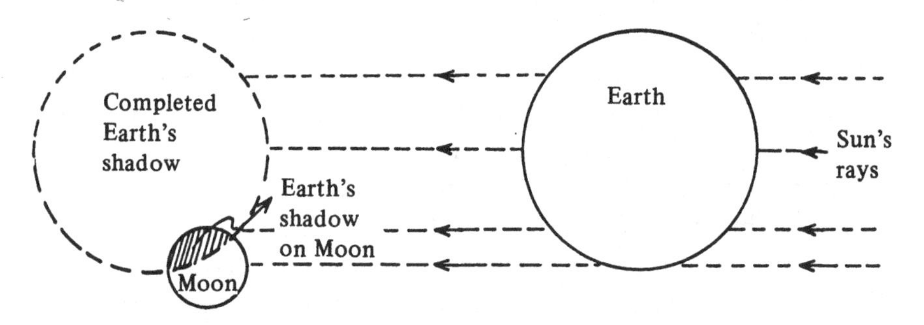


Fig. 2. By observing the shadow of the Earth on the Moon and completing the observed arc to form a complete circle, we can see that the ratio of the Moon's diameter to the Earth's diameter is the ratio of the circle's (representing the Moon) diameter to the completed circle's diameter. This is roughly 1:4.

500 miles is the ratio of 360° to 7°. From here it is easy to see that Eratosthenes obtained the very good estimate of about 25,000 miles. This estimate was used shortly afterwards by Hipparchus to estimate the size of the Moon. The Moon was observed during a partial lunar eclipse. The arc of the Earth's shadow on the Moon could be extended to form a complete circle and the ratio of the diameter of the Moon's disc to the shadow of the Earth's disc would be the ratio of their actual diameters (see Fig. 2). The estimate was close to 2,000 miles for the Moon's diameter – again an excellent estimate. Since the angular diameter of the Moon was known to be

$\frac{1}{2}^\circ$, the distance to the Moon could be easily worked out (see Fig. 3) to be roughly 250,000 miles – once again an excellent estimate. The next step was to note that during a solar eclipse the Moon just covers the disc of the Sun. Thus the angular diameter of the Sun is also $\frac{1}{2}^\circ$. By observing the angle made at the Earth by the Moon and the Sun at half-moon (see Fig. 4), knowing the distance from the Earth to the Moon, the distance to the Sun and hence the size of the Sun could be estimated.

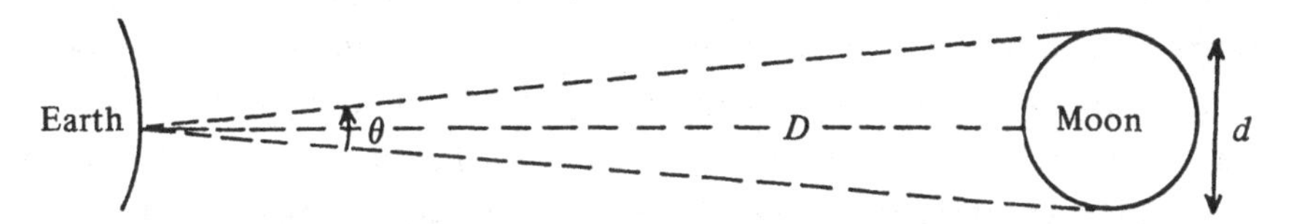


Fig. 3. If the angular diameter of the Moon as seen on Earth, is $\theta$ (measured in radians) the distance to the Moon, $D$, is $d\theta$, where $d$ is the Moon's diameter.

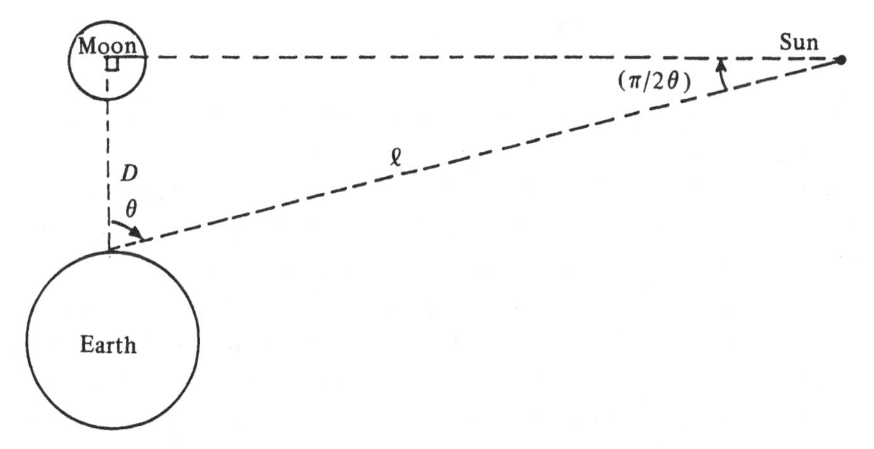


Fig. 4. When there is a half-Moon the Sun-Moon-observer on Earth forms a right angle triangle. Observing the angle, $\theta = \widehat{MES}$, and knowing $D$, the distance $ME$, we can work out $\ell = D/\cos\theta$, the distance $ES$.

These estimates were available to Ibn-al-Haytham, over a thousand years later. He revived the view of Aristarchus. If on no other count, then just the sheer size of the Sun would have convinced him that the Earth went round the Sun. He showed that the planets moved in circles around the Sun. Two centuries later, Al Zarkali modified these results, in the light of better data, to state that they moved in ellipses with the Sun at one focus. In the meantime Al Kindi stated his 'law of terrestrial gravity: "All terrestrial objects are attracted towards the centre of the Earth". Thus the 'Arabs' modified the Aristotilean laws of celestial and terrestrial motion.

For completeness, it may be mentioned that the ancients knew of six planets: the Moon, the Sun, Venus, Mars, Jupiter, Saturn and the fixed stars. Before Aristotle the beliefs appear to be that there were 7 separate moving domes. The Arabs discovered Mercury and thus provided 7 planets. The fact that Mercury is so obviously a satellite of the Sun may have played a significant role in convincing Ibn-al-Haytham that all the planets (except the Moon) are satellites of the Sun. By this stage, Muslim civilisation was on the decline and the next developments in the study of motion came from the West.

## 2. The Measurement of Time

With hindsight we can say that one of the main problems in the study of motion has been the measurement of spatial and temporal intervals. As regards the measurement of space intervals, they were soon refined enough for the purpose. Certainly, the ancient Greeks were able to obtain very accurate measures of distance. However, the measurement of time remained a problem for long afterwards. The subjective impression of the passage of time was as inaccurate as that of distance or temperature or other such quantities. The problem was the lack of an objective measure that was fine enough to study motion. There were available crude measures such as the day, the month and the year. Early on these were broken into four parts, and later even more parts. However none of these were refined enough to study terrestrial motion. The first real 'clocks' were the sun-dial and the water clock. The sun-dial consisted of a dial with an upright piece which cast a shadow. Since the rate at which the shadow moved would depend on the time of day and the season, and could be seen only during the day and when the sky was clear, this 'clock' was not very reliable. The water-clock consisted of a bucket with a hole near the bottom and a marking near the top. It would be filled up to the marking and then allowed to empty out. The time taken of course depended on the heat and humidity, which would determine the loss of water not through the hole, but due to evaporation. Also, the hole would expand at higher temperatures, allowing more water through. In all, it was much too crude a measure of time to be useful. In fact, there was (and remains) the notion that there are 'different times' since there was no reliable time-measure which was significantly better than the subjective impression of the passage of time.

Fig. 5. A schematic representation of an hour-glass. The ends are closed.

The first major improvement in the technology of time measurement comes with the production of smooth, clear, glass moulded into different shapes. In the West, where the next developments arose, the 'hour glass' came into use towards the start of the Middle Ages. It consists of glass moulded in the form shown in Fig. 5 with sand trickling from one end to the other, both ends being closed. After the sand passes from one end completely, the glass is turned upside down. For the first time, with this development, a time interval of about an hour could be measured objectively distinctly better than it could be measured subjectively. Presumably the hour glass was invented in Arabia, which seems the most likely to develop the technology used for its construction. However, even this development was not adequate for the proper study of motion. The Middle Ages do, nevertheless, mark the start of the scientific study of terrestrial motion, as we shall see later.

Galileo Galilei, the discoverer of so many other important facts and principles, was the first to provide a time-measuring device for sufficiently small intervals of time. He noticed that what we now call a pendulum swings with the same period regardless of the amplitude of the oscillation. He verified this belief by timing the swing against his pulse. Further, he noticed that a shorter pendulum swings faster than a longer one. Thus he was able to construct 'clocks' which measured different intervals and could be calibrated against each other. He himself used such clocks to measure the rate of motion in given situations. From this empirical work he drew certain conclusions which could be stated as the first modern laws of motion.

Since then there have been further improvements in the technology of time measurement. The spring watch was the earlier one and the electronic and atomic clocks the more recent. The atomic clock essentially measures time by 'counting the number of electromagnetic waves' of a given wavelength emitted by a particular element. The current accuracy of time measurement is about $10^{-19}$ sec! It should be borne in mind that such clocks were not available to Einstein. At the start of the development of Relativity, the accuracy was only about $10^{-1}$ sec.

It should be clear already that after the start of the Renaissance there was no reason for confusion about the concept of time. As pointed out earlier, the subjective assessment of time is unreliable at best, as with the subjective assessment of distance, force, temperature, humidity, etc. At worst, as in dreams, it can be entirely misleading. Unfortunately, many philosophers, over the generations, have continued to mystify the time-concept. The validity of their hair-splitting arguments is doubtful, but there can be no doubt of their lack of relevance for practical purposes. There are very good objective means of measuring time. However, they are not necessarily equivalent. It is necessary, when talking of time measurement, to specify the means of measurement. For example, if we define time as being measured by a pendulum clock, the clock will slow down as we go higher above the surface of the Earth. We would have to conclude that 'time dilates' as we go up. For the purposes of the present work it is adequate to take the time concept as that defined by measurement by atomic clocks (even though they have become available only recently and were *not* available for the study of motion being discussed, at that time).

## 3. Classical Motion

The first major breakthrough in the study of motion was when Baron Simon disproved Aristotle's belief that, in general, heavier bodies fall faster than lighter ones. He made the following arrangement (see Fig. 6). A wooden board was placed over a large hollow in the ground to provide a sound-box such as one has in musical

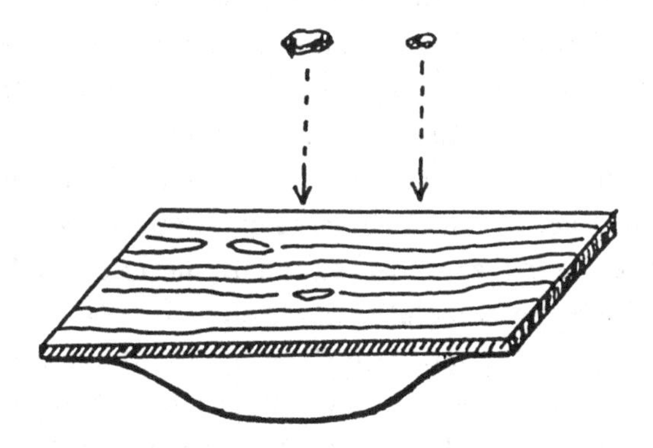

Fig. 6. A schematic representation of the demonstration that heavier and lighter objects fall at the same rate. The hollow beneath the wooden plank makes an enormous sound-box which can identify any beats produced. The null result of this experiment proves that there is no significant time lag between the falling of the smaller and the larger stone.

instruments (particularly of the stringed variety). From a second storey he arranged to drop two stones simultaneously. He found that, regardless of whether the stones were of equal weight or one was much heavier than the other, no beats were produced. If the stones had fallen at different times beats would have been produced. Thus they fell equally fast.

Galileo used his invention of the pendulum clock to study, quantitatively, how bodies moved on the Earth. He found that an object once put in motion would continue in its uniform motion if it were not for friction or other forces acting on it. By performing experiments on an inclined plane he inferred that the height through which a body falls is proportional to the square of the time it takes to fall, unless other forces act on it.

There were also major advances made in the study of celestial motion. Nicolai Copernicus, a Polish monk, had revived the views of Ibn-al-Haytham. According to this view, Mercury, Venus, Earth, Mars, Jupiter and Saturn followed concentric, circular orbits of increasing radius about the Sun. The Moon followed a circular path about the Earth. Beyond Saturn were the fixed stars. (This picture is nowadays known as the Copernican system instead of Aristarchus' or Ibn-al-Haytham's system.)

Again Galileo played a major role in this study. He developed and improved the recent invention of the telescope. He used it, not merely as a toy for looking at distant objects on the Earth, but as a scientific instrument to study the Heavens. He deduced that the shadows on the face of the Moon were due to mountains. By carefully measuring the shadow as a ratio of the diameter of the Moon and the angle of the Sun, he calculated the height of the lunar mountains. In fact he mapped the entire surface of the Moon very accurately. He also discovered four satellites of Jupiter and studied their motion carefully. His work clearly marked the end of the Aristotilean laws of celestial motion. His discovery of 'novae' (new stars) was the death knell of the belief that the Heavens are eternal and perfect. With him the so-called Copernican system came into its own. Not that everybody accepted his views and findings immediately. Quite the contrary. There was an uproar each time he announced results at variance with Aristotle's beliefs and he was repeatedly forced to recant those views. Nevertheless, his findings gained currency – fairly quickly for those times.

The telescope and Galileo's findings were used to good effect by Tycho Brahe to collect data. The data he collected was used by Johannes Kepler. Kepler revived Al Zarkali's law of planetary motion, which states that planets move in ellipses with the Sun at one focus. (This is now known as Kepler's first law rather than Al Zarkali's law.) He went on, however, to state two more laws which were quantitative. These laws were vital for the further development of the study of the motion of celestial objects.

Then came Newton! He used Galileo's law, which he restated as follows: "Every body continues in its state of rest or of uniform motion unless an external force acts on it". This statement is nowadays known as Newton's first law of motion. The second law was stated as: "The rate of change of the amount of motion is proportional to the force causing the change". The 'amount of motion', as distinct from the 'rate of motion', was *momentum* rather than *speed*. A consequence of these laws was his third law of motion: "Every action has an equal and opposite reaction". This was not enough to explain motion in general. Robert Hooke had earlier quantified and modified Al Kindi's law of terrestrial gravity for explaining celestial motion to state that: "All objects are pulled towards the Sun with a force proportional to their mass and inversely proportional to the square of their distance from the Sun". Thus he had stated a law of celestial gravity. This law led to Kepler's laws as a consequence. Newton generalised this law to the law of universal gravity: "Every body attracts every other body with a force proportional to the product of their masses and inversely proportional to the square of the distance between them". He thus managed to unify motion in the Heavens with that on Earth. *The same laws apply everywhere in the Universe.* For completeness, it should be added that Newton thought of light as composed of corpuscles possessing mass and hence he expected that the path of light would be bent by a gravitational source. He also talked of the paths of planets being 'refracted' about the Sun. This is a remarkable insight presaging the General Theory of Relativity.

The opinion on the nature of light that was generally accepted was that of Christian Huyghens. He believed that light was a form of energy propagated by wave motion. Of course, the motion had to be in some medium. For this purpose Aristotle's 'aether' was modified to serve as a medium in which travelled wave-like disturbances that we call light. Later many different 'aethers' were required. The medium for light was called the 'luminiferous aether'. The 'aether' also provided a frame for absolute rest.

The study of both wave motion and usual mechanical motion continued. Most notable were the contributions of Lagrange and Hamilton in this respect. The essential problem they tackled was of celestial mechanics. Since every body attracted every other body, not only did the Sun attract each of the planets, but the planets also attracted the Sun. In fact they also attracted each other. The procedure originally adopted was to solve the problem for the Sun with each planet separately and then apply corrections for each of the other planets. The corrected result would be applied to provide further corrections, and so on. Later Lagrange developed a method of dealing with all ten bodies together on the same footing using generalised coordinates and velocities to express the total free-energy of the system and minimise it. Then Hamilton used generalised coordinates and momenta to express the total energy of the system. These are the methods of Lagrangian and Hamiltonian mechanics that are used so extensively nowadays. We will need to refer to them later.

In the meantime there had been extensive investigations into the phenomena of electricity and magnetism. In the nineteenth century these phenomena were unified in Maxwell's theory of electromagnetism. He developed a set of equations to describe these phenomena. In addition, he showed that there were electromagnetic waves which would travel in a vacuum with the speed of light and in dielectric media with a correspondingly slower speed. The conclusion was quite unavoidable – light is an electromagnetic wave.

The 'electromagnetic aether' is the 'luminiferous aether'. By about 1880 Newton's theory as augmented by Maxwell's work seemed complete. Maxwell had introduced the concept of a field, which was the tendency to influence a test particle (sufficiently small so as not to influence the field). Such a field could be visualised as the lines of force about a bar magnet that can be traced by placing iron filings near it. This concept has, since then, become very important due to its application in modern physics where it often replaces the generalised coordinates in the Lagrangian or the Hamiltonian (the free or the total energy).

The stage is now set to present the events which led to the formulation of Relativity.

## 4. Pre-Relativistic Mechanics

Towards the end of the nineteenth century, Lorentz tried to complete the theory of electromagnetism by including discrete charged particles. There were some basic problems involved but the $\beta$-rays of J. J. Thompson had been identified as streams of charged particles, nowadays called 'electrons'. A theory for the motion of these particles was necessary. Lorentz published it in the form of a book entitled 'The Theory of Electrons'. In order to make the theory self-consistent he had to introduce certain ad-hoc assumptions. One was that the 'electromagnetic mass', i.e., the mass relevant for the theory, was velocity-dependent according to the formula

$$m_{em} = \frac{m}{\sqrt{1 - v^2/c^2}} \ . \tag{1.1}$$

The other was that there must be a transformation of coordinates and a new 'local time' parameter had to be introduced:

$$\left.\begin{array}{l} x' = \gamma\varepsilon(x - vt), \quad y' = y, \quad z' = z, \quad t' = \gamma\varepsilon(t - vx/c^2)\,, \\ \gamma = \frac{1}{\sqrt{1-v^2/c^2}} \ . \end{array}\right\} \tag{1.2}$$

In the mean time Poincaré had been discussing the theory of motion from a more philosophical and mathematical point of view.

He stated categorically that only *relative* motion should be discussed, that there was no meaning to be attached to absolute motion. He called this the principle of relativity of motion.

The most significant development was Michaelson's attempt to measure the velocity of Earth through the aether. If a measurement was made at a time when the rotation of the Earth (at the place of measurement) was in the same direction as the revolution of the Earth about the Sun, the two velocities should add. Thus, if the Sun was at rest in the aether there would be a speed of about 30 km/sec. of the Earth through the aether. If the Sun was moving then at some time of the year, when the Earth's motion was in the same direction as the Sun's, the speed would be even greater. He showed that with his newly developed interferometer he should be able to measure such a velocity accurately and thus indirectly 'see the aether'. The basis for this expectation is contained in the following argument.

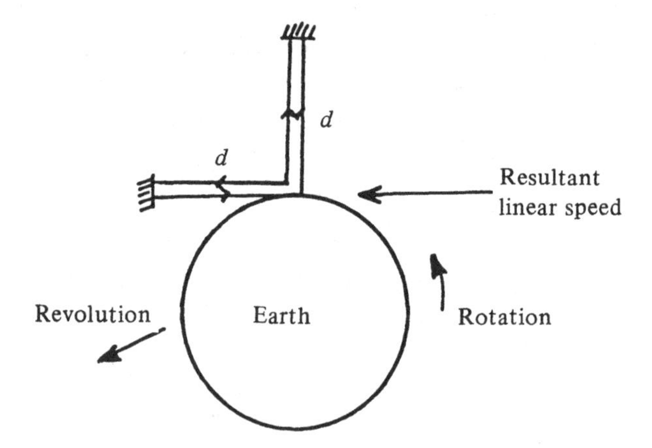


Fig. 7. The essentials of the Michaelson-Morley experiment. Light is sent along the direction of motion of the Earth, (classically) at a speed $(c + v)$ and back against the motion (classically) at a speed $(c - v)$. Perpendicular to the motion the speed of both legs of the journey must be the same. The time-lag could be measured by seeing interference fringe-shifts, but none were seen.

Consider two rays of light sent out from a point on the Earth's surface, one along the direction of motion and one perpendicular to it. Let both be reflected back, as shown in Fig. 7. For convenience, suppose that the distance travelled in both directions is the same: $d$. If the speed of light is denoted by $c$ and the speed of the Earth through the aether by $v$, we can work out the time taken by the light to travel in each direction. Along the direction of motion it is

$$t_1 = \frac{d}{c+v} + \frac{d}{c-v} = \frac{2d/c}{1 - v^2/c^2} \ . \tag{1.3}$$

To travel perpendicular to the direction of motion it must be directed into the effective aether wind (see Fig. 8) so that the ***resultant*** velocity is perpendicular to the direction of motion. As is clear from Fig. 8, the magnitude of the resultant velocity of light is

$$c_r = c\sqrt{1 - v^2/c^2} \ . \tag{1.4}$$

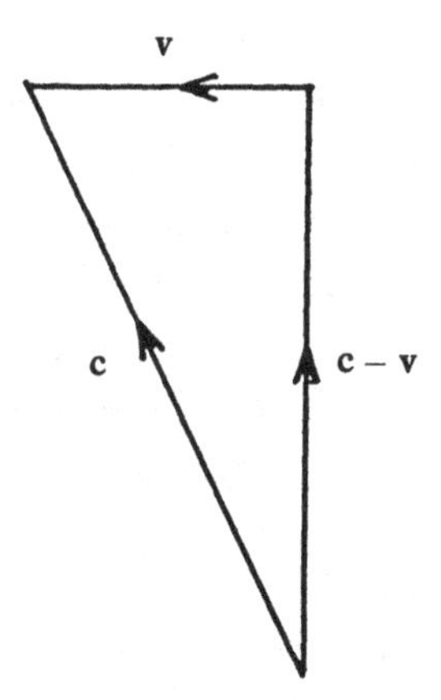


Fig. 8. If light is sent with a velocity **c** so that the resultant velocity, (**c** − **v**) is perpendicular to $v$, the magnitude of this resultant vector is $\sqrt{c^2 - v^2}$.

Thus, the time taken by this ray of light is

$$t_2 = \frac{2d/c}{\sqrt{1 - v^2/c^2}} \ . \tag{1.5}$$

The time lag, $\Delta t$, is then

$$\Delta t = t_1 - t_2 = \frac{2d/c}{1 - v^2/c^2}\left(1 - \sqrt{1 - v^2/c^2}\right) . \tag{1.6}$$

Thus, we get

$$\frac{\Delta t}{t_1} = 1 - \sqrt{1 - v^2/c^2} \approx \frac{1}{2}\frac{v^2}{c^2} , \tag{1.7}$$

for sufficiently small $v/c$. In this case, we expect $v/c \approx 10^{-4}$. Thus, $2\Delta t/t_1 \approx 10^{-8}$. Now, if the situation is changed so that the ray that was perpendicular to the direction of motion now goes against it and back, and the ray that went in the direction of motion now goes perpendicular, we would get a total effect of $2\Delta t \approx 10^{-8} \times t_1$.

Michaelson developed an instument which split a beam of light into two parts which could travel different distances, and so the waves comprising the light beams would arrive out of phase with each other. They would then 'interfere' with each other 'constructively' at some places and 'destructively' at others. This instrument is called a Michaelson interferometer. Now a change of the speed of light would lead to a phase shift corresponding to a change in the positions of constructive and destructive interference. Where there is constructive interference there is a bright band and where there is destructive interference a dark band. These bands are called 'interference fringes'. The change of the speed of light should, therefore, lead to a fringe shift. Since the wavelength of light is very small, phase shifts of extremely small duration can be detected. It is found that for the expected motion of the Earth there should be a significant shift in the interference fringes.

Michaelson's attempt was significant because it failed. Despite repeated attempts with many precautions taken and all sorts of improvement made in collaboration with Morley, it failed. Not a fraction of a fringe shifted. It was as if the fringes were painted onto the eye-piece, so definitely fixed they remained. By 1895 there was no room for doubt that the Michaelson-Morley *null* result was valid. An explanation of this result was required. It was suggested that maybe the aether 'dragged' along with the Earth so that there was no relative motion between the Earth and the aether. However, if it did it should take some energy of motion from the Earth and hence reduce the Earth's energy causing it to spiral in towards the Sun. This did not happen. To avoid this conclusion we must postulate that the aether took no energy because it was massless. In that case it would be impossible to 'drag' it as any force would accelerate it infinitely. The aether already had been postulated to be infinitely hard. This property was required to explain the fact that the speed of light in a vacuum was greater than in any medium. The essence of the argument is as follows. A wave travels faster in a hard than in a soft medium. If there is aether everywhere where there is no matter, but is displaced by matter, and it is harder than any substance, we would expect light to travel faster where there is less matter than where there is more matter. It might seem odd that the aether did not stop the Earth's motion through it, but that could be explained away by postulating that it passes through matter with no resistance. However, this makes the 'dragging of aether' even more implausible.

An alternative 'explanation' was suggested by Fitzgerald. Suppose, he said, that because of some unknown dynamical process, there is a contraction of length in the direction of motion for all physical bodies, but there is no such effect in the directions perpendicular to the motion. Let this shortening be given by

$$d' = d\sqrt{1 - v^2/c^2} \ . \tag{1.8}$$

Now, in Eq. (1.3) $d'$ would appear instead of $d$. Thus $t_1$ would be equal to $t_2$ and so no time-lag could be expected. Independently, but somewhat later, Lorentz showed that if we take $\varepsilon = 1$ in Eqs. (1.1) and (1.2), Eq. (1.8) follows from there. He also believed that there would be some dynamic, or electrodynamic, process which would account for this contraction, now known as the Lorentz-Fitzgerald contraction.

## 5. A Digression on Scientific Method

Before continuing with the study of motion it is necessary to discuss (very briefly) what is meant by 'science', 'scientific theory' and 'scientific method'. In ancient times they referred to the attempt to provide causes for observed phenomena. Since these were assigned to the whims of some unseen personalities or personality, mythology and religion formed the basis of 'science'. With Aristotle, 'science' became the process of finding the 'self-evident truth' which explained observed phenomena. In the Muslim civilisation 'science' became essentially the collection and systematic collation of data by observation. With Galileo, the data collection began to include experimentation in a modern sense. With Newton it became the reduction of all phenomena to mechanical models by which they could be understood in terms of the 'laws of Nature' discovered by Newton. (A more detailed discussion of this subject is available in my article 'Modern Scientific Thought in Perspective' in *The History of Science in Central Asia*, ed. A. Qadir, Centre for the Study of Central Asian Civilisations, Quaid-i-Azam University Press, 1975.)

The modern view, expressed by Karl Popper, is largely based on Einstein's work in Quantum Theory and Relativity. This is not accepted unanimously, but it is important to grasp fully the theory of Relativity. According to this view, 'doing science' means following the 'scientific method' to comprehend phenomena. The 'scientific method' consists of formulating 'scientific theories' which explain all known phenomena and then testing them. A 'scientific theory' is a set of assumptions which leads to a falsifiable conclusion, i.e., one which could *in principle* be proved wrong. To be able to test a prediction quantitatively there must be a procedure to measure the relevant quantities explicitly. In addition, there is an infinite set of assumptions which provide the concepts used. The collection of all these gives us a *physical theory*. (Details are given in A. Qadir, *Int. J. Theoret. Phys.* **15** (1976) 635–641.) It is necessary that any 'explanation' of observed phenomena is a physical theory.

## Exercise 1

1. A man walks to work at a speed of 6km/hr. How accurately would it be necessary to measure his walking stick to be able to detect the Lorentz-Fitzgerald contraction of the stick if it is 1 m long? Is such an accuracy physically attainable?

2. An electronic device capable of measuring changes of $10^{-2}$ cm is being used at a fencing match, and it sees a 1 m foil decreased in size. Considering the matter quantitatively, could the decrease be due to the Lorentz-Fitzgerald contraction?

3. What are the percentage changes in length due to the Lorentz-Fitzgerald contraction for the following?
   (a) A train.
   (b) A racing car.
   (c) A jet plane.
   (d) A satellite.
   (e) A deep-space reconaissance vehicle (like Mariner).
   (f) The Earth moving round the Sun.

4. Is Fitzgerald's suggestion a physical theory? Does Lorentz's suggestion improve Fitzgerald's idea or make it worse?

5. If light is effected by gravitation, it should be possible for it to go into orbit (i.e., a closed path) about a gravitational source. How dense would a thousand kilogramme mass have to be for light to be in a circular orbit about it?

# Chapter 2

# DERIVATION OF SPECIAL RELATIVITY

## 1. Einstein's Formulation of Special Relativity

In 1905 Einstein solved the problems of the day by appealing to kinematics rather than dynamics. It was because of the lack of any dynamic reasoning that his theory met with such strong resistance initially. He first analysed classical kinematics and showed that it would lead to an observer-dependent speed of light. He then showed that this result would lead to stellar aberration which was not observed. The prediction was quite independent of the Michaelson-Morley null experiment. (However, we shall use the latter experiment in our discussion.) He then showed how, if we assume that the speed of light is observer-independent, the Lorentz transformations follow as a consequence. Further, he showed that the 'local time parameter' must be treated as a genuine, physical, time.

The classical kinematic transformations for uniform linear motion with speed $v$, which Einstein called the Galilean transformations, are

$$x' = x - vt, \;\; y' = y, \;\; z' = z, \;\; t' = t, \tag{2.1}$$

where motion is in the $x$-direction only. Differentiating these equations with respect to $t'$, bearing in mind that $t' = t$, gives the formula for the resultant velocity,

$$u'_x = u_x - v, \;\; u'_y = u_y, \;\; u'_z = u_z \;\; . \tag{2.2}$$

Taking $\mathbf{u} = (c,0,0)$ we see that $\mathbf{u}' = (c - v,0,0)$. Thus, if we take $v$ as negative the magnitude of $\mathbf{u}'$, $|\mathbf{u}'|$, is greater than $c$. This result would lead to various problems with stellar aberration and the theory of electromagnetism and electrons. Most important of all, it leads to the supposedly measurable time-lag that was not found by Michaelson and Morley.

Einstein now made explicit two assumptions. On these assumptions he based his explanation. The assumptions made were the following:

(a) *The principle of special relativity*, that all inertial frames are physically equivalent;

(b) *The principle of the constancy of the speed of light*, that the speed of light in vacuum (approximately $3 \times 10^8$ m/sec) is constant for all inertial observers.

An 'inertial frame' is a frame of reference in which Newton's second law of motion holds. On the Earth, for example, it does not hold because an object without support falls instead of continuing in its state of rest. Here, the external force acting on it is gravity. In an inertial frame an object without support should stay in its place nevertheless. Principle (a) states that there is no physical difference for any two observers in inertial frames even if they move relative to each other in that 'physical laws' appear the same to both observers. This way the absence of absolute motion gets reformulated. Principle (b) simply says that the speed of light is independent of the speed of the observer though it could depend on accelerations, etc. Clearly the theory is restricted to dealing with uniform (linear) motion. It is for this reason that it was called the Restricted, or Special, Theory of Relativity. Einstein spent 10 more years formulating a workable Non-restricted, or General, Theory of Relativity to deal with arbitrary motions. Here we will not follow Einstein's original derivation of his result, or his subsequent derivations. Rather, I would like to present a procedure for derivation which fits in with a more general and simpler formulation of the axioms underlying the theory. These will be stated after deriving Einstein's basic results.

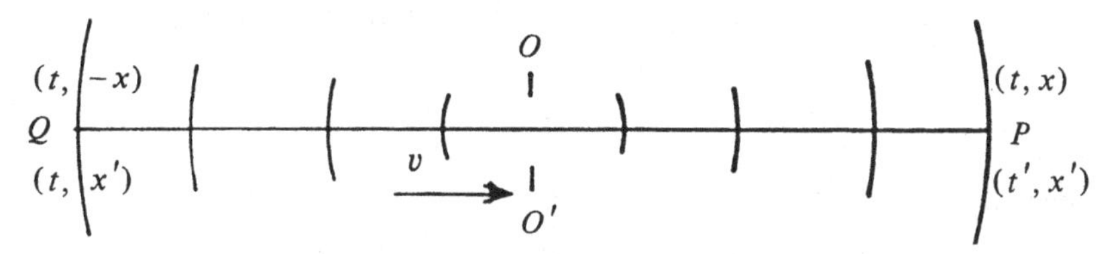


Fig. 9. The thought experiment of Special Relativity. Light signals (or signals at maximum speed), sent by relatively moving observers $O$ and $O'$ must travel together. This enables us to provide the moving wave-front ($P$ and $Q$) to translate from one frame to another.

Consider two observers $O$ and $O'$, where $O'$ is moving with speed $v$ in the $x$-direction relative to $O$. Thus, of course, $O$ is moving with speed $-v$ relative to $O'$. At one instant the two observers 'coincide'. By this, we mean that there is some small (negligible) displacement between them in a direction perpendicular to the $x$-direction, but there is no displacement in the $x$-direction. They both take this instant as their origin of time, i.e., at this instant they start their clocks. They both also send two light signals in the positive and negative $x$-directions (see Fig. 9). Now, because the speed of light is the same for all observers, the signals travel together. Let $O$ measure time and space by the coordinates $(t, x, y, z)$ and $O'$ by $(t', x', y', z')$. To translate the coordinates of one into the other we need to have an agreed point and determine how the coordinates of one are used in terms of the coordinates of the other. Let the signal, at any instant, be at $P$ in the positive direction and $Q$ in the negative direction. The equations for $P$ and $Q$, respectively, according to $O$, are

$$ct - x = 0 \ (P), \quad ct + x = 0 \ (Q) \ , \tag{2.3}$$

and according to $O'$ are

$$ct' - x' = 0 \ (P), \quad ct' + x' = 0 \ (Q) \ . \tag{2.4}$$

Since $P$ is given by both equations, each implies the other. For the reciprocity implicit in principle (a) to hold,

$$ct' - x' = \lambda(ct - x) \ , \tag{2.5}$$

where $\lambda$ is a constant of proportionality. Similarly, for $Q$

$$ct' + x' = \mu(ct + x) \ , \tag{2.6}$$

where $\mu$ is a constant of proportionality.

Adding Eqs. (2.5) and (2.6) and dividing by 2,

$$ct' = act - bx \ , \tag{2.7}$$

where $a$ and $b$ are defined by

$$a = (\lambda + \mu)/2, \ \ b = (\lambda - \mu)/2 \ . \tag{2.8}$$

Subtracting Eq. (2.5) from Eq. (2.6) and dividing by 2 gives

$$x' = -bct + ax \ . \tag{2.9}$$

We now need to determine $a$ and $b$.

To determine $b$, notice that the equation for the position, $x$, of $O'$ according to $O$ is

$$x = vt \ . \tag{2.10}$$

According to $O'$ it is, of course, $x' = 0$. Putting these values into Eq. (2.9) gives us

$$0 = -bct + avt \ . \tag{2.11}$$

Since this equation holds for all $t$, we have

$$b = av/c \ . \tag{2.12}$$

Thus Eqs. (2.7) and (2.9) become

$$ct' = a(ct - \frac{v}{c}x) \ , \tag{2.13}$$

$$x' = a(x - \frac{v}{c}ct) \ . \tag{2.14}$$

To determine $a$ we have to appeal to principle (a). It may be rephrased to say that on interchanging the primed and unprimed indices there should be no difference (except for a change of $v$ to $-v$). Now, let us define $x_0$ as $x$ at $t = 0$ and $x_0'$ as $x'$ at $t' = 0$. Putting $t = 0$ in Eq. (2.14) gives

$$x_0/x' = 1/a \ . \tag{2.15}$$

To obtain an expression for $x_0'$ we need to obtain a relation between $x$ and $t$ when $t' = 0$ and use that relation to obtain an expression for $x'$ at $t' = 0$. Putting $t' = 0$ in Eq. (2.13)

$$ct\big|_{t'=0} = \frac{v}{c}\,x\big|_{t'=0} \ . \tag{2.16}$$

Inserting this relationship in Eq. (2.14) gives us

$$x_0' = a(x - \frac{v}{c}\frac{v}{c}x) \ . \tag{2.17}$$

Thus, we get

$$x_0'/x = a(1 - v^2/c^2) \ . \tag{2.18}$$

Now, by principle (a)

$$x_0/x' = x_0'/x \ . \tag{2.19}$$

Using Eqs. (2.15), (2.18) and (2.19) we see that

$$a = \frac{1}{\sqrt{1 - v^2/c^2}} \ . \tag{2.20}$$

There has been no effect due to motion in the direction perpendicular to the motion. Thus, Eqs. (2.13), (2.14) and (2.20) are in fact the Lorentz transformations, Eqs. (1.2), with $a = \gamma$ and $\varepsilon = 1$. The Lorentz transformations are purely kinematic and have no need to appeal to dynamics or electrodynamics. Before drawing any further conclusions from here, let us look at the simpler statement of the theory.

### 2. Reformulation of Relativity

A question could arise whether the speed of light is the same for all observers, or that it changes so little from observer to observer that we are unable to detect that change with our present technology. In other words 'how reliable is Relativity theory?' and 'how strongly does it depend on experimental data?'. It turns out that we do not even really need to assume principle (b). We could construct the set of all physically attainable speeds and without loss of generality call its least upper bound $C$. If the set is not bounded from above, we would have an infinite value of $C$. Now, we could repeat the previous argument with signals sent at this maximum speed $C$. Since it is a maximum speed it is the same for all observers. This may be seen simply by noticing that the signal in the positive $x$-direction sent by $O$ cannot reach any point *after* that sent by $O'$, as it is travelling at the fastest possible speed, and vice versa. All we have to do then is replace $c$ by $C$.

Another modification is that we do not need to restrict ourselves to inertial frames only, even for principle (a). All we really need here is that they be relatively unaccelerated frames. In one sense this is a weaker formulation, because the assumption is stronger. On the other hand the theory gets wider applicability since it *can* deal with non-inertial frames as well.

In this reformulation we will need one experiment to determine $C$ and another to test the theory. In the usual formulation one experiment to test the theory would have been adequate. It turns out that within the limits of experimental error $C = c$, and we *do* have other predictions to test the theory with. We shall therefore use $c$ throughout. Even if it turned out that $C$ is not *exactly* equal to $c$ it would not change the theory much; we would just use $C$ instead of $c$. We now proceed to derive the basic consequences of the theory.

## 3. Length Contraction, Time Dilation and Simultaneity

The Lorentz transformations are not directly *physically* testable. They refer to coordinates only. For testing we need to have predictions in terms of intervals. We could, for example, consider what happens to time intervals. Let there be a time interval $\delta t$ as seen by $O$. By this we mean that there were two times when somebody in the frame of $O$ looked at his clock, $t_1$ and $t_2$. The position remained the same according to $O$. Thus, $x_1 = x_2$. Let us write down the Lorentz transformations for both points $(t_1, x_1)$ and $(t_2, x_2)$:

$$t_1' = \gamma\big(t_1 - \frac{v}{c^2}\, x_1\big), \quad t_2' = \gamma\big(t_2 - \frac{v}{c^2}\, x_2\big) \ , \tag{2.21}$$

where we have, by definition,

$$\delta t = t_1 - t_2, \quad \delta t' = t_1' - t_2' \ . \tag{2.22}$$

Thus,

$$\delta t' = \gamma\Big[(t_1 - t_2) - \frac{v}{c^2}\,(x_1 - x_2)\Big] \ . \tag{2.23}$$

Since $x_1 = x_2$, we see that

$$\delta t' = \gamma \delta t = \frac{\delta t}{\sqrt{1 - v^2/c^2}} \ . \tag{2.24}$$

This is known as the *time dilation formula*. It implies that the unit of time measurement of $O'$ is larger than that of $O$. Thus, in the same interval fewer of the units of $O'$ will fit in than those of $O$. Thus the clocks of $O'$ will appear to run slow.

Another possibility is to measure spatial intervals. Let the spatial interval according to $O$ be

$$\delta x = x_1 - x_2 \ . \tag{2.25}$$

Now $O'$ must see the two ends of the interval at the same time *according to him*, i.e., $t_1' = t_2'$. From Eq. (2.13) we see that then

$$t_1 - t_2 = \frac{v}{c^2}\,(x_1 - x_2) \ . \tag{2.26}$$

Again using the Lorentz transformations

$$\begin{aligned} \delta x' = x_1' - x_2' &= \gamma\Big((x_1 - x_2) - \frac{v^2}{c^2}\,(x_1 - x_2)\Big) \\ &= \gamma \cdot \delta x \cdot (1 - v^2/c^2) \ . \end{aligned} \tag{2.27}$$

Thus, we get the Lorentz-Fitzgerald length contraction

$$\delta x' = \delta x/\gamma = \delta x\sqrt{1 - v^2/c^2} \ . \tag{2.28}$$

Both these formulae have been thoroughly tested: the first by flying atomic clocks in a jet plane and comparing them with other atomic clocks in the laboratory, the second by radar tracking spacecraft. As mentioned earlier, $c$ here has the same value as our best estimates of the speed of light in vacuum.

Consider now two events that appear simultaneous to $O$, i.e., one occurs at $x_1$ and the other at $x_2$ at the same time $t$, or in other words $t_1 = t_2 = t$. According to $O'$, they occur at

$$t'_1 = \gamma(t_1 - \frac{v}{c^2}\, x_1) = \gamma(t - \frac{v}{c^2}\, x_1) \ , \tag{2.29}$$

and at

$$t'_2 = \gamma(t_2 - \frac{v}{c^2}\, x_2) = \gamma(t - \frac{v}{c^2}\, x_2) \ . \tag{2.30}$$

Thus, we see that $t'_1 \neq t'_2$ as

$$t'_1 - t'_2 = \gamma\, \frac{v}{c^2}\, (x_2 - x_1) \ . \tag{2.31}$$

Hence ***simultaneity is relative.*** It is the relativity of simultaneity, rather than of motion, that gives the theory its name. It should be pointed out at this stage that a common error is the belief that according to this theory (i.e., Relativity) ***everything*** is relative. This is just not true. In fact the theory attempts to formulate that part of physical theory that can be stated in absolute terms. It is only motion (known classically) and simultaneity (stated by Einstein) that are taken to be relative. ***Some*** other quantities turn out to be relative as well.

Many supposed 'paradoxes' were constructed to try to 'disprove' Relativity. In fact, they merely highlighted misconceptions that can arise due to mixing the older concepts with the results of Relativity. The most famous one – the clock (or twin) paradox – will be discussed later. Here I would like to discuss one which arises from

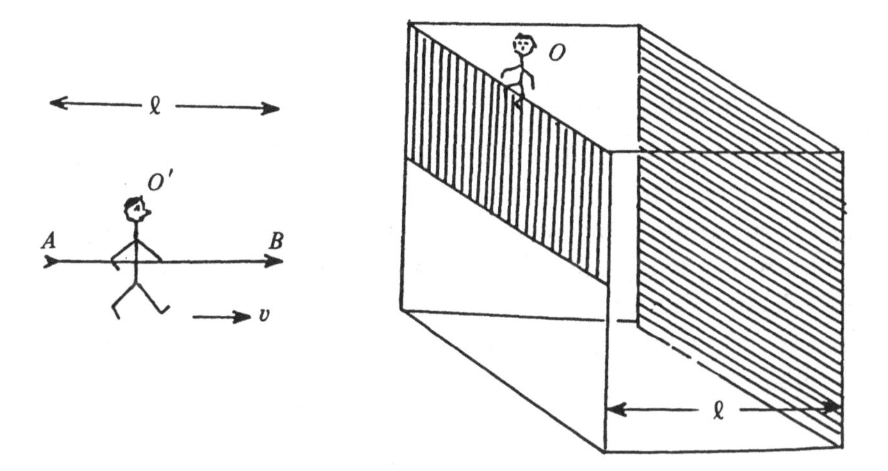


Fig. 10. Observer $O$ sits tending a door that can be dropped behind observer $O'$ who comes running into the barn of rest-length $\ell$ with a spear $AB$, also of rest-length $\ell$.

not adverting to the relativity of simultaneity. Imagine an observer $O'$ running relative to an observer $O$ with a speed $v$ (see Fig. 10). $O'$ carries a spear, $AB$, of length $\ell$, while $O$ sits on a barn having a length $\ell$, towards which $O'$ is running. $O$ can control a trap-door which he can close instantaneously. Now, according to $O$, the length of $AB$ is not $\ell$ but $\ell' = \ell/\gamma$. Thus, $O$ reasons, he can close the door with $AB$ inside the barn. On the other hand, $O'$ reasons that the barn has a length $\ell' = \ell/\gamma$. Hence $O$ cannot close the door with $AB$ inside the barn, as it has length $\ell > \ell'$. The question is, which one is correct?. The answer is that both are correct! How? Well, when we say that $AB$ is in the barn we do not mean that it is forever in the barn. If it maintains its speed it must break through the barn at the other end. Thus, what we are really saying is that *at some instant* it is in the barn, i.e., both ends $A$ and $B$ are *simultaneously* in the barn. Stated this way the problem is automatically resolved since simultaneity is relative.

We will discuss the so-called clock 'paradox' later, along with some other paradoxes which arise due to the presence of a gravitational field or some accelerational effect. They are of a fundamentally different nature to the above 'paradox'.

## 4. The Velocity Addition Formulae

The whole argument, so far, began with the observation that the Galilean transformations yield an observer-dependent speed of light. The basic requirement, as far as we are concerned, is that $c$ (or $C$) is observer-independent. Thus, we should obtain a different formula for velocity addition, by which $c$ is independent of $v$. To obtain the required formulae we write the Lorentz transformations, Eqs. (1.2), for differentials

$$dt' = \gamma(dt - v dx/c^2) \ , \tag{2.32}$$

$$dx' = \gamma(dx - v dt) \ , \tag{2.33}$$

$$dy' = dy \ , \tag{2.34}$$

$$dz' = dz \ . \tag{2.35}$$

Now the definition of the speed of any object in the $x$, $y$ and $z$ directions according to $O$, is

$$u_x = dx/dt, \ u_y = dy/dt, \ u_z = dz/dt \ . \tag{2.36}$$

According to $O'$ the corresponding speeds must be defined by

$$u'_x = dx'/dt', \ u'_y = dy'/dt', \ u'_z = dz'/dt' \ . \tag{2.37}$$

Thus, dividing Eqs. (2.33), (2.34) and (2.35), respectively, by Eq. (2.32), we get

$$u'_x = \frac{u_x - v}{1 - u_x v/c^2} \ , \tag{2.38}$$

$$u'_y = \frac{u_y\sqrt{1 - v^2/c^2}}{1 - u_x v/c^2} \ , \tag{2.39}$$

$$u'_z = \frac{u_z\sqrt{1 - v^2/c^2}}{1 - u_x v/c^2} \ . \tag{2.40}$$

These are the required velocity addition formulae which give the correct physical procedure to put velocities together. If, for example, $\mathbf{u} = (c, 0, 0)$, we get, as required

$$u'_x = \frac{c - v}{1 - cv/c^2} = \frac{c - v}{1 - v/c} = c \ . \tag{2.41}$$

## 5. Three Dimensional Lorentz Transformations

The previous analysis depended on the assumption that the motion was entirely in the $x$-direction. This assumption could be justified by arguing that we could define the $x$-direction as the direction of motion, since it would be a matter of choice. However, it is not always convenient to define directions in that way. We may want to use polar coordinates or to generalise to non-uniform motion. For this reason, it would be useful to develop the Lorentz transformations for motion in any arbitrary direction.

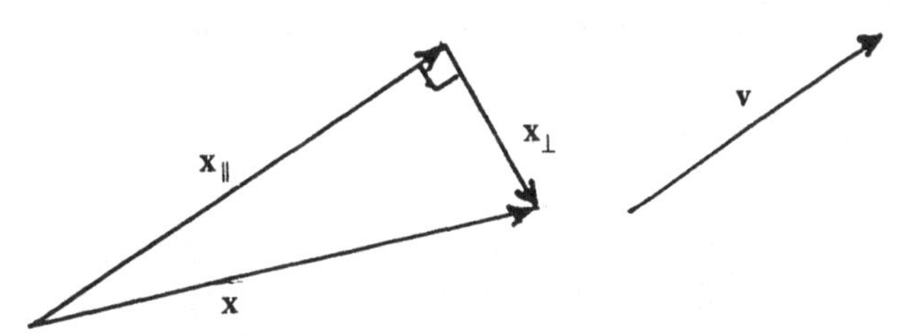


Fig. 11. The break-up of $\mathbf{x}$ into $\mathbf{x}_{\parallel}$ along $\mathbf{v}$ and $\mathbf{x}_{\perp}$ orthogonal to $\mathbf{v}$.

To state the general Lorentz transformations we note that we can split the position vector, $\mathbf{x}$, into two parts – one along the direction of motion ($\mathbf{x}_{\parallel}$) and the other perpendicular to it ($\mathbf{x}_{\perp}$), as shown in Fig. 11,

$$\mathbf{x} = \mathbf{x}_{\parallel} + \mathbf{x}_{\perp} \ . \tag{2.42}$$

Now it is clear that there is no effect of uniform motion on $\mathbf{x}_{\perp}$, i.e.,

$$\mathbf{x}'_{\perp} = \mathbf{x}_{\perp} \ . \tag{2.43}$$

However $\mathbf{x}_{\parallel}$ must transform as the $x$-coordinate transforms

$$\mathbf{x}'_{\parallel} = \gamma(\mathbf{x}_{\parallel} - \mathbf{v}t) \ , \tag{2.44}$$

where the Lorentz factor, $\gamma$, must obviously be given by

$$\gamma = (1 - \mathbf{v}\cdot\mathbf{v}/c^2)^{-\frac{1}{2}} \ . \tag{2.45}$$

Since $\mathbf{x}_{\parallel} \propto \mathbf{v}$ and $\mathbf{x}\cdot\mathbf{v} = \mathbf{x}_{\parallel}\cdot\mathbf{v}$, it is easily seen that

$$\mathbf{x}_{\parallel} = (\mathbf{x}\cdot\mathbf{v}/\mathbf{v}\cdot\mathbf{v})\mathbf{v} \ . \tag{2.46}$$

From Eq. (2.42), we have

$$\mathbf{x}_{\perp} = \mathbf{x} - \mathbf{x}_{\parallel} = \mathbf{x} - \mathbf{v}(\mathbf{x}\cdot\mathbf{v}/\mathbf{v}\cdot\mathbf{v}) \ . \tag{2.47}$$

Adding Eqs. (2.43) and (2.44) we see that

$$\begin{aligned} \mathbf{x}' &= \mathbf{x}'_{\parallel} + \mathbf{x}'_{\perp} \\ &= \gamma(\mathbf{x}_{\parallel} - \mathbf{v}t) + \mathbf{x}_{\perp} \ . \end{aligned}$$

Thus we get the Lorentz transformations for general motion

$$t' = \gamma(t - \mathbf{v}\cdot\mathbf{x}/c^2) \ , \tag{2.48}$$

$$\mathbf{x}' = \mathbf{x} + \mathbf{v}\big[(\mathbf{x}\cdot\mathbf{v}/\mathbf{v}\cdot\mathbf{v})(\gamma - 1) - \gamma t\big] \ . \tag{2.49}$$

Equation (2.49) can be used to derive the general formula corresponding to the length contraction formula. (Notice that the time dilation formula remains unchanged, except that $\gamma$ is given by Eq. (2.45) now.) Consider the general spatial vector

$$\mathbf{r} = \mathbf{x}_1 - \mathbf{x}_2 \ . \tag{2.50}$$

We want to determine $\mathbf{r}'$ as seen by $O'$, i.e., with $t'_1 = t'_2$. Now, by Eq. (2.48) we see that

$$t_1 - t_2 = \mathbf{v}\cdot\mathbf{x}_1/c^2 - \mathbf{v}\cdot\mathbf{x}_2/c^2 = \mathbf{v}\cdot\mathbf{r}/c^2 \ . \tag{2.51}$$

Using Eqs. (2.50) and (2.51) with (2.49) we obtain

$$\mathbf{r}' = \mathbf{r} + \mathbf{v}\left(\frac{\mathbf{r}\cdot\mathbf{v}}{\mathbf{v}\cdot\mathbf{v}}\right)\left(\gamma - 1 - \gamma\frac{\mathbf{v}\cdot\mathbf{v}}{c^2}\right) \ , \tag{2.52}$$

which is the generalised 'length contraction formula'. It reduces to

$$\mathbf{r}' = \mathbf{r} + \mathbf{v}(\mathbf{r} \cdot \mathbf{v}/\mathbf{v} \cdot \mathbf{v})(\gamma^{-1} - 1) \ . \tag{2.53}$$

To obtain the general velocity addition formulae, we follow the previous procedure of writing the relevant equations, Eqs. (2.48) and (2.49), as differentials and dividing the latter by the former. By definition

$$\mathbf{u} = \mathrm{d}\mathbf{x}/\mathrm{d}t, \ \mathbf{u}' = \mathrm{d}\mathbf{x}'/\mathrm{d}t' \ . \tag{2.54}$$

Thus the general formulae become

$$\mathbf{u}' = \frac{\mathbf{u} + \mathbf{v}[(\mathbf{u} \cdot \mathbf{v}/\mathbf{v} \cdot \mathbf{v})(\gamma - 1) - \gamma]}{\gamma(1 - \mathbf{u} \cdot \mathbf{v}/c^2)} \ . \tag{2.55}$$

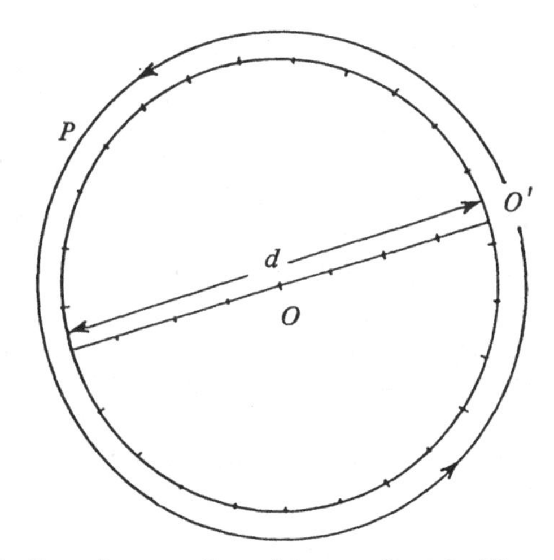


Fig. 12. The rotating observer thought experiment. The rotating observer sees the rods along the circumference shortened, but not those along the diameter.

We shall apply this discussion to consider the instantaneous Lorentz transformations for an observer in uniform circular motion. Of course, this motion is accelerated. However, if we consider the frame of reference which is inertial and at the instant has the same motion as the rotating frame, we can meaningfully talk of Lorentz transformations. Einstein has already shown (see *The Meaning of*

*Relativity* by A. Einstein) that in this frame the geometry is non-Euclidean. He proved it by the following thought experiment (see Fig. 12). Consider a large circle with metre rods along its diameter and making up its circumference. Let the circumference as measured by these rods be $p$ and the diameter be $d$. Then, by Euclidean geometry we know that $p/d = \pi$. Now, let the circle of rods be rotated. Due to the motion, an observer at rest would perceive the rods along the circumference to be shortened. This is due to the Lorentz-Fitzgerald contraction. However, the rods along the diameter would be unaffected as they would be orthogonal to the direction of motion. Thus, instead of $p$ the perimeter would appear to be $p\gamma^{-1}$. Hence the ratio of circumference to diameter for the moving circle, as seen by the observer at rest, would be $p'/d' = p/\gamma d = \pi/\gamma$. Hence the geometry would be non-Euclidean.

To derive the Lorentz transformations notice that $\mathbf{v}\cdot\mathbf{x} = 0$. Thus Eqs. (2.48) and (2.49) for a frame in uniform circular motion are

$$t' = \gamma t, \ \mathbf{x}' = \mathbf{x} - \gamma \mathbf{v} t \ . \tag{2.56}$$

Here, if the angular speed is $\Omega$, $\gamma = (1 - r^2\Omega^2/c^2)^{-\frac{1}{2}}$ where the radius of the circle is $r$. In cylindrical coordinates we get

$$t' = \frac{t}{\sqrt{1 - r^2\Omega^2/c^2}} \ , \ r' = r \ , \ \theta' = \frac{\theta - \Omega t}{\sqrt{1 - r^2\Omega^2/c^2}} \ , \ z' = z \ . \tag{2.57}$$

Many textbooks give incorrect Lorentz transformations for this case.

## Exercise 2

1. An observer, $A$, sees a body as having twice the length that another observer, $B$, sees. Which of them has the greater speed relative to the body if it lies along the direction of their relative motion? If a third observer sees the length as three times that seen by $B$ and the body is in the rest-frame of one of the three observers, which is the rest-frame of the body?

2. Determine the speed with which one object must move relative to another for its clock to be slowed by 1% as seen by the other.

3. Two observers, $A$ and $B$, see a rocket taking off at distances $d$ and $d'$ at times $t$ and $t'$ respectively, having set their clocks at zero when they were coincident. If their direction of relative motion is in line with the rocket, find the speed of $B$ relative to $A$.

4. Two observers, $A$ and $B$, see the same event. $A$ sees it at a distance of $3 \times 10^7$ km at exactly 10 o'clock by his watch, while $B$ sees it at a distance of $2 \times 10^7$ km at 5 minutes and 10 seconds to 10 by his watch. If the time when $B$ passed by $A$ was exactly 10 to 10 by $A$'s watch and exactly a quarter to 10 by $B$'s watch, and the event was in the direction of relative motion, determine the speed of $B$ relative to $A$.

5. Particles of half-life $10^{-8}$ secs are produced 3 km above sea level and most of them are found at sea level. What is the least speed at which they must be travelling? In their rest-frame there is no time dilation. How is it that they, nevertheless, can travel 3 km?

6. A scientific satellite finds some cosmic radiation to be entirely composed of neutrons (which have a half-life of about 1,000 sec) coming at $(1 - 10^{-7})c$. Could the radiation have been produced outside the solar system (about $10^{10}$ km across) or must it originate from within the solar system?

7. Prove that the relativistic resultant of three co-linear speeds $u$, $v$, $w$ is given by

$$\frac{u + v + w + uvw/c^2}{1 + (uv + vw + wu)/c^2} .$$

What will be the formula for $n$ co-linear speeds when $n$ is even and when $n$ is odd?

8. An observer sees a clock as showing 1 hour to be half an hour. If he sees an object lying at an angle of $\pi/4$ as having a length of 2 m, what is the rest-length of the object?

9. An observer, $O$, sees a rod as having a length 1 m and making an angle $\pi/4$ with the direction of motion. If, in its rest-frame, $O'$, the rod makes an angle $\pi/6$ with the direction of motion, what is the rest-length of the rod?

10. An observer, $O$, sees three bodies ($A$, $B$ and $C$) having velocities $(u, 0, 0)$, $(0, -v, 0)$, $(0, 0, w)$ respectively. What is the velocity of each body relative to the other? In particular, is the velocity of $A$ as seen by $B$ exactly opposite in direction and equal in magnitude to the velocity of $B$ as seen by $A$? If not, why is there any difference? Would the velocities be equal and opposite if the velocity of $B$ were $(0, v, 0)$ instead of $(0, -v, 0)$?

11. An observer, $A$, sees another observer, $B$, as moving with a velocity $(u, v, 0)$ and $B$ sees $C$ moving with a velocity of $(u, 0, v)$. What is the speed of $C$ relative to $A$?

12. $P$ sees $Q$ moving with a velocity $(u, 0, 0)$ and $Q$ sees $R$ moving with a velocity $(0, v, 0)$ while $R$ sees $S$ moving with a velocity $(0, 0, w)$. If a metre rod makes an angle of $\pi/6$ with the z-axis in the yz-plane in the frame of $S$, what will its length appear to be according to $A$?

13. Three spaceships $A$, $B$, $C$ are engaged in battle. $A$ is a scout ship of the fleet to which $B$ belongs and $C$ is an enemy ship. $A$ tells $B$ that $C$ is in its direct line of flight at a distance of 3,000 km coming at a speed of 108,000 km/h towards it. If $B$ is behind $A$ at a distance of 1,000 km and moving in the opposite direction with a velocity of 36,000 km/h relative to $A$, where should $B$ aim a laser beam? How big must the ship be for him to be able to ignore: (a) the time-lag for the information being sent; and (b) relativistic effects?

14. An observer, $O$, sees a metre rod at an angle of $\pi/4$ to the x-axis as being $\sqrt{5/6}$. Another observer, $O'$, sees the metre rod as a metre rod and the clock on a body, $B$, running at half speed. What is the maximum speed of $B$ relative to $O$ and what is its minimum speed? Why is the velocity not uniquely given? What is the apparent angle of the rod according to $O$?

15. An observer, $O'$, sees a body at a distance $\ell$ as having an angular speed $\Omega$ and no other velocity component, while $O$ sees $O'$ as moving radially outward with a speed $v$. Determine the velocity of the body as seen by $O$.

16. A rod of length $\ell$ makes angles $\pi/3$ and $\pi/4$ with the x- and y-axes respectively. What does the length of the rod appear to be to an observer moving with a velocity $(c/\sqrt{3},\ c/\sqrt{6},\ c/\sqrt{3})$?

17. A spaceship sends out a scout ship with a velocity $(c/2,\ c/3,\ c/4)$. The scout ship spies an enemy ship approaching with a velocity $(c/4,\ c/3,\ c/2)$. What velocity should the scout ship tell its mother ship that the enemy ship is approaching at?

18. The length of a moving object appears to be halved while a clock on it appears to be running three times too slow. What is the angle between the object and its direction of motion in its own rest-frame?

# Chapter 3

# DIGRESSION INTO TENSOR THEORY

## 1. Invariant Quantities

We have been considering the transformation of some quantities due to some actual physical processes. We found that some quantities constructed from them were invariant under the mathematical transformations representing these physical processes. For a more powerful formulation of Special Relativity it is necessary to obtain a more formal understanding of these invariant quantities. A *scalar* quantity is invariant under coordinate transformations. For example, the magnitude of a 2-dimensional vector, $\mathbf{a}$, is a scalar. It remains unchanged by translating the origin, rotating the axes through any angle or converting to plane polar coordinates. In fact the magnitude of any vector, in any space, is an invariant quantity, being a scalar. Of course, the position vector of a point does not satisfy this requirement as it is defined not only by the point but also by the origin.

A *vector* is also an invariant quantity. This statement may seem strange after discussing Lorentz transformations, which seem to give the transformation of a position 4-vector, without changing the origin. The point is that the quantity being transformed is *not* the vector but its components. In usual 3-dimensional space any vector field, i.e., a position dependent vector $\mathbf{A}(\mathbf{x})$, can be written as

$$\mathbf{A}(\mathbf{x}) = A^1(\mathbf{x})\mathbf{e}_1 + A^2(\mathbf{x})\mathbf{e}_2 + A^3(\mathbf{x})\mathbf{e}_3 \ , \tag{3.1}$$

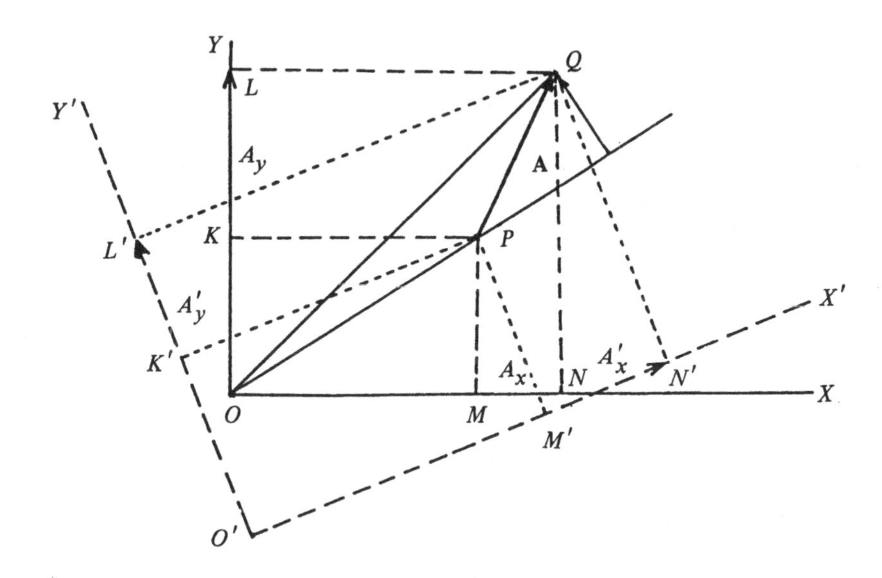


Fig. 13. By transforming coordinates $\mathbf{A} = \overrightarrow{PQ}$ is not altered. However, it has components $(A_x, A_y)$ in $XOY$ and $(A'_x, A'_y)$ in $X'O'Y'$. Since the coordinates of $P$ and $Q$ transform by the same matrix so does $\overrightarrow{PQ} = \mathbf{A}$, in its components.

where $\mathbf{e}_1$, $\mathbf{e}_2$, $\mathbf{e}_3$ are the three basis vectors in the given coordinate system and $A^1(\mathbf{x})$, $A^2(\mathbf{x})$, $A^3(\mathbf{x})$ are the corresponding components of the vector. The entire set of basis vectors may be more compactly represented by $\mathbf{e}_i$, bearing in mind that $i$ can take the values 1, 2, 3. Similarly the components can be represented more compactly by $A^i(\mathbf{x})$. Then Eq. (3.1) may be rewritten in the form

$$\mathbf{A}(\mathbf{x}) = \sum_{i=1}^{3} A^i(\mathbf{x})\mathbf{e}_i \ . \tag{3.2}$$

Einstein introduced a further simplification of notation by taking the *summation convention*, that repeated upper and lower indices are summed over. We will generally follow this convention, explicitly stating any deviation from it. Using this convention, Eq. (3.2) becomes

$$\mathbf{A}(\mathbf{x}) = A^i(\mathbf{x})\mathbf{e}_i \ . \tag{3.3}$$

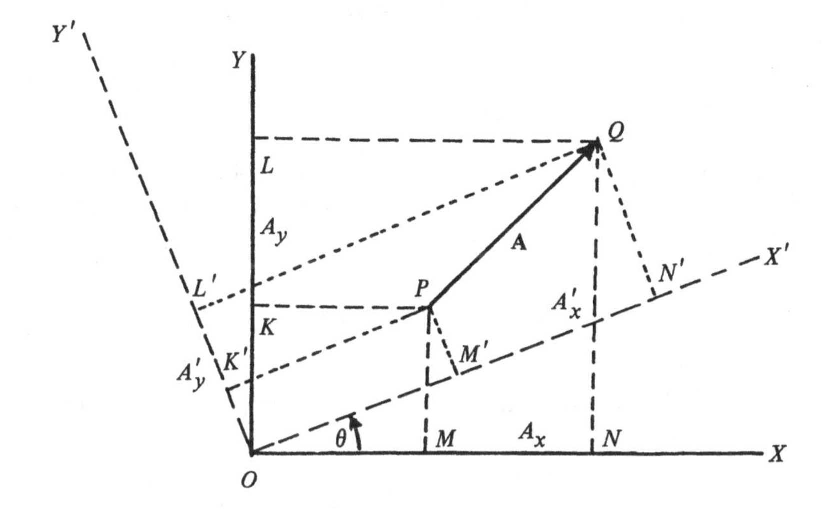


Fig. 14. The vector $\mathbf{A} = \overrightarrow{PQ}$, has Cartesian components in the $XOY$-frame $A_x = MN$, $A_y = KL$ and in the $X'O'Y'$-frame, $A'_x = M'N'$, $A'_y = K'L'$. It can be shown that $M'N' = MN\cos\theta + KL\sin\theta$, $K'L' = KL\cos\theta - MN\sin\theta$.

As seen from Fig. 13, the vector **A** is invariant under transformations of the coordinate system, but its components, $A^i$, do vary. Clearly, if there is a rotation in the XY-plane, $A^1$ and $A^2$ are transformed by the usual rotation matrix (see Fig. 14). Thus, for a rotation through an angle $\theta$, the transformation is given by

$$\begin{pmatrix} A'^1 \\ A'^2 \\ A'^3 \end{pmatrix} = \begin{pmatrix} \cos\theta & \sin\theta & 0 \\ -\sin\theta & \cos\theta & 0 \\ 0 & 0 & 1 \end{pmatrix} \begin{pmatrix} A^1 \\ A^2 \\ A^3 \end{pmatrix} . \tag{3.4}$$

This matrix also gives the transformation of coordinates.

How can we explain this transformation in terms of Eq. (3.3)? The point is that $\mathbf{e}_i$ are the basis vectors in one coordinate system and will change to $\mathbf{e}'_i$ in the transformed coordinate system. Also, $A^i$ will change to $A'^i$ in such a way that

$$A^i(x)\mathbf{e}_i(x) = A'^a(x')\mathbf{e}'_a(x') . \tag{3.5}$$

Now, for general coordinate transformations the generalization of Eq. (3.4) is

$$A'^a(x') = \frac{\partial x'^a}{\partial x^i} A^i(x) \ . \tag{3.6}$$

A vector satisfying the transformation law, given by Eq. (3.6), for its components, is called a ***contravariant vector***. Inserting this equation into Eq. (3.5), we see that

$$A^i(x)\Big(\mathbf{e}_i(x) - \frac{\partial x'^a}{\partial x^i}\, \mathbf{e}'_a(x')\Big) = 0 \ . \tag{3.7}$$

Since Eq. (3.7) holds for all $A^i(x)$, the expression inside the brackets must be zero. Hence

$$\mathbf{e}_i(x) = \frac{\partial x'^a}{\partial x^i}\, \mathbf{e}'_a(x') \ . \tag{3.8}$$

Now notice that in the Einstein summation convention

$$\frac{\partial x^i}{\partial x'^b} \frac{\partial x'^a}{\partial x^i} = \left(\frac{\partial x^i}{\partial x'^b} \frac{\partial}{\partial x^i}\right) x'^a = \frac{\partial x'^a}{\partial x'^b} = \delta^a_b \ , \tag{3.9}$$

the Kronecker delta defined by $\delta^a_b = 1$ if $a = b$, $\delta^a_b = 0$ if $a \neq b$. Thus the Kronecker delta is the identity matrix and merely substitutes a new index for a previous one, i.e.,

$$\delta^a_b\, A^b = A^a, \ \ \delta^a_b\, B_a = B_b \ . \tag{3.10}$$

Multiplying Eq. (3.8) by $\partial x^i/\partial x'^b$ and using Eqs. (3.9) and (3.10) we see that

$$\mathbf{e}'_b(x') = \frac{\partial x^i}{\partial x'^b}\, \mathbf{e}_i(x) \ . \tag{3.11}$$

A ***covariant vector***, $\mathbf{B}(\mathbf{x})$, is defined by

$$\mathbf{B}(\mathbf{x}) = B_i(x)\mathbf{e}^i(x) \ . \tag{3.12}$$

From the foregoing discussion it is clear that

$$B'_a(x') = \frac{\partial x^i}{\partial x'^a} B_i(x) \tag{3.13}$$

and

$$\mathbf{e}'^{a}(x') = \frac{\partial x'^{a}}{\partial x^{i}} \mathbf{e}^{i}(x) \ . \tag{3.14}$$

The scalar or dot product of two vectors can only be defined if one is covariant and the other is contravariant so that the two transformations cancel out and leave the product invariant. Using Eqs. (3.6), (3.9) and (3.13), we see that

$$\begin{aligned} A'^{a}(x')B'_{a}(x') &= \frac{\partial x'^{a}}{\partial x^{i}} \frac{\partial x^{j}}{\partial x'^{a}} A^{i}(x)B_{j}(x) \\ &= \delta_{i}^{j} A^{i}(x)B_{j}(x) = A^{i}(x)B_{i}(x) \ . \end{aligned} \tag{3.15}$$

To be able to define a scalar product of two contravariant or two covariant vectors we need to convert one type of vector into the other.

## 2. Tensors

Generally, invariant quantities are called *tensors*. Their invariance is assured in much the same way as the invariance of vectors, namely the variation of one part is cancelled by the variation of another part. Thus, for example, a scalar obtained by taking the scalar product of two vectors is a tensor. Of course, so are the covariant and the contravariant vectors. Another example would be the tensor

$$\mathbf{A} = A^{ij}(x)\mathbf{e}_{i}(x)\mathbf{e}_{j}(x) \ , \tag{3.16}$$

provided its components satisfy the transformation rule

$$A'^{ab}(x') = \frac{\partial x'^{a}}{\partial x^{i}} \frac{\partial x'^{b}}{\partial x^{j}} A^{ij}(x) \ , \tag{3.17}$$

so as to cancel the variation of $\mathbf{e}_i$ and $\mathbf{e}_j$ with coordinate transformations. A tensor whose components satisfy Eq. (3.17) is called a *contravariant tensor*. Similarly there can be a *covariant tensor*

$$\mathbf{B} = B_{ij}(x)\mathbf{e}^{i}(x)\mathbf{e}^{j}(x) \ , \tag{3.18}$$

whose components satisfy the covariant transformation law

$$B'_{ab}(x') = \frac{\partial x^i}{\partial x'^a} \frac{\partial x^j}{\partial x'^b} B_{ij}(x) \ . \tag{3.19}$$

Also, there can be a ***mixed tensor*** with one part contravariant and the other covariant,

$$\mathbf{C} = C^i_j(x)\mathbf{e}_i(x)\mathbf{e}^j(x) \ , \tag{3.20}$$

whose components satisfy a mixed transformation law

$$C'^a_b(x') = \frac{\partial x'^a}{\partial x^i} \frac{\partial x^j}{\partial x'^b} C^i_j(x) \ . \tag{3.21}$$

Notice that the terms in the above expressions can be put in any order since the index label contains the information of how matrix multiplication is to proceed. However, the order of the indices is fixed. In general, $A^{ij} \neq A^{ji}$ for example.

It is clear that tensors can be constructed using any number of basis vectors, e, together. The number of e's involved in a tensor is called its *rank*. The nature of the tensor depends on how many of the e's have lower indices (like $\mathbf{e}_i$) and how many upper indices (like $\mathbf{e}^i$). If the number of lower index e's is $\ell$ and the number of upper index e's is $k$, the tensor is said to be of *valence* $\begin{pmatrix} k \\ \ell \end{pmatrix}$. Clearly the rank of the tensor will be $(k+\ell)$. If $\ell = 0$ the tensor is contravariant, if $k = 0$ it is covariant and if neither is zero, it is mixed. Thus, a scalar is a tensor of rank zero, a vector is a tensor of rank one, and tensors of rank more than one are the new quantities defined.

Of particular interest is the *metric tensor*, g, which is a covariant tensor that associates to every contravariant vector a unique covariant vector, thus enabling us to define the scalar product of two contravariant vectors. The *length* of a vector is the square root of the dot product of the vector with itself. Thus, the metric tensor enables us to define the length of a contravariant vector – hence its name. Generally, we have

$$A_k = g_{kj}\, A^j = g_{jk}\, A^j \ , \tag{3.22}$$

where all the components may generally be position dependent. To be able to associate back the contravariant vector with the covariant vector, we require that $g_{ij}$ be invertible, i.e., g must be invertible. Thus, there must exist $\mathbf{g}^{-1}$ such that

$$\mathbf{g}^{-1} \cdot \mathbf{g} = \mathbf{g} \cdot \mathbf{g}^{-1} = \mathbf{I} \ , \tag{3.23}$$

where $\mathbf{I}$ is the identity tensor, with components $\delta^i_j$, i.e., it is the identity matrix in any coordinate system. It is clear that $\mathbf{g}^{-1}$ must be a contravariant tensor. Its components are generally written as $g^{ij}$, so that Eq. (3.23) can be rewritten as

$$g^{ik}\, g_{jk} = \delta^i_j \ . \tag{3.24}$$

Thus, multiplying Eq. (3.22), by $g^{ik}$ gives, using Eq. (3.23),

$$g^{ik}\, A_k = g^{ik}\, g_{jk}\, A^j = \delta^i_j\, A^j = A^i \ , \tag{3.25}$$

giving us back the original contravariant vector.

Now, given a vector using Cartesian coordinates,

$$\mathbf{A} = A^x\, \mathbf{i} + A^y\, \mathbf{j} = A^i\, \mathbf{e}_i \ , \tag{3.26}$$

we know that the square of its length is given by

$$\begin{aligned} A^2 &= (A^x)^2 + (A^y)^2 = A_i A^i = g_{ij}\, A^i A^j \\ &= g_{11}\, (A^x)^2 + 2g_{12}\, A^x A^y + g_{22}\, (A^y)^2 \ . \end{aligned} \tag{3.27}$$

By direct comparison we see that in this case $g_{11} = 1 = g_{22}, g_{12} = 0$. Generally, with Cartesian coordinates $g_{ij} = 1$ if $i = j$ and $g_{ij} = 0$ if $i \neq j$. Thus, it behaves much like the identity matrix, except that it is entirely covariant. If we represent $A^i$ as a column vector and, for consistency $A_i$ as a row vector, then $g_{ij}$ is a partitioned row matrix. For example, in two dimensions,

$$g_{ij} = (1 \quad 0 \ \vdots \ 0 \quad 1) \ , \tag{3.28}$$

so that

$$A_i = (1 \quad 0 \ \vdots \ 0 \quad 1) \begin{pmatrix} A^x \\ A^y \end{pmatrix} = (A^x \quad A^y) \ . \tag{3.29}$$

In other words $g_{ij}$ simply transposes the vector. It is clear from here why there is no difference between covariant and contravariant vectors in Cartesian coordinates. However, this is not the case when other coordinate systems are used.

## 3. Coordinate Transformations

So far we have dealt with coordinate transformations in a very abstract way. To make the procedure more concrete we shall consider some specific examples. To start with we consider the transformation from Cartesian to plane polar coordinates. In this case

$$x^1 = x, \ x^2 = y; \ x'^1 = r, \ x'^2 = \theta \ , \tag{3.30}$$

where the two sets of coordinates are related by

$$\left.\begin{aligned} &x = r\cos\theta, \ \ y = r\sin\theta \ , \\ &r = \sqrt{x^2+y^2} \ , \ \ \theta = \tan^{-1}(y/x) \ . \end{aligned}\right\} \tag{3.31}$$

Now the coordinate transformation matrix is given by

$$\left(\frac{\partial x'^a}{\partial x^i}\right) = \begin{pmatrix} \partial r/\partial x & \partial r/\partial y \\ \partial\theta/\partial x & \partial\theta/\partial y \end{pmatrix} \ , \tag{3.32}$$

and its inverse transformation is given by

$$\left(\frac{\partial x^i}{\partial x'^a}\right) = \begin{pmatrix} \partial x/\partial r & \partial x/\partial\theta \\ \partial y/\partial r & \partial y/\partial\theta \end{pmatrix} \ . \tag{3.33}$$

Now, it is easily seen from Eq. (3.31) that

$$\left.\begin{aligned} &\frac{\partial r}{\partial x} = \frac{x}{r} = \cos\theta, \ \frac{\partial r}{\partial y} = \frac{y}{r} = \sin\theta \ , \\ &\frac{\partial\theta}{\partial x} = -\frac{y/x^2}{1+\tan^2\theta} = -\frac{\sin\theta}{r}, \ \frac{\partial\theta}{\partial y} = \frac{1/x}{1+\tan^2\theta} = \frac{\cos\theta}{r} \ . \end{aligned}\right\} \tag{3.34}$$

Also, we have

$$\left.\begin{array}{l} \dfrac{\partial x}{\partial r} = \cos\theta, \quad \dfrac{\partial x}{\partial \theta} = -r\sin\theta \ , \\ \dfrac{\partial y}{\partial r} = \sin\theta, \quad \dfrac{\partial y}{\partial \theta} = r\cos\theta \ . \end{array}\right\} \tag{3.35}$$

Thus, the coordinate transformation matrix and its inverse are given by

$$\begin{aligned} \left(\frac{\partial x'^a}{\partial x^i}\right) &= \begin{pmatrix} \cos\theta & \sin\theta \\ -\sin\theta/r & \cos\theta/r \end{pmatrix} ; \\ \left(\frac{\partial x^i}{\partial x'^a}\right) &= \begin{pmatrix} \cos\theta & -r\sin\theta \\ \sin\theta & r\cos\theta \end{pmatrix} . \end{aligned} \tag{3.36}$$

It is easy to check, from Eq. (3.36), that as required

$$\left(\frac{\partial x^i}{\partial x'^a}\right) = \left(\frac{\partial x'^a}{\partial x^i}\right)^{-1} . \tag{3.37}$$

Consider the vector **A** with contravariant Cartesian components $A^x$ and $A^y$, so that its covariant components $A_x = A^x$ and $A_y = A^y$. Now using Eqs. (3.6) and (3.36), we have

$$\left.\begin{aligned} A^r &= A'^1 = A^1 \cos\theta + A^2 \sin\theta \\ &= A^x \cos\theta + A^y \sin\theta \ , \\ A^\theta &= A'^2 = A^1(\sin\theta/r) + A^2(\cos\theta/r) \\ &= (A^y \cos\theta - A^x \sin\theta)/r \ . \end{aligned}\right\} \tag{3.38}$$

This transformation is shown diagrammatically in Fig. 15.

Now let us see the transformation of the covariant components using Eqs. (3.13) and (3.36) to give

$$\left.\begin{aligned} A_r &= A'_1 = A_1 \cos\theta + A_2 \sin\theta = A^x \cos\theta + A^y \sin\theta \ , \\ A_\theta &= A'_2 = A_1(-r\sin\theta) + A_2 r\cos\theta = r(A^y \cos\theta - A^x \sin\theta) \ , \end{aligned}\right\} \tag{3.39}$$

so that $A_r = A^r$ and $A_\theta = r^2 A^\theta \neq A^\theta$! Here we see a difference between contravariant and covariant components.

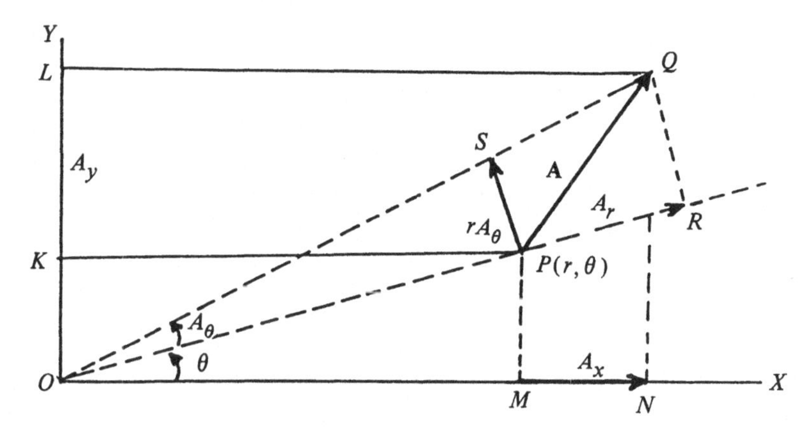


Fig. 15. The polar and Cartesian components of the vector $\mathbf{A}\cdot A_x = \overline{MN}$, $A_y = \overline{KL}$, $A_r = \overline{PR}$. Since the polar coordinates of $P$ are $(r,\theta)$, we see that $PS = rA_\theta$. Also $A_x = A_r \cos\theta$, $A_y = A_r \sin\theta$.

Let us also consider the transformation of the metric tensor from Cartesian to plane polar coordinates. Remember that $g_{11} = 1 = g_{22}$, $g_{12} = 0 = g_{21}$. Since **g** is a covariant tensor, Eq. (3.19) is applicable to it. Hence we have

$$\left.\begin{aligned}
g'_{11}(x') &= \left(\frac{\partial x^1}{\partial x'^1}\right)^2 \cdot 1 + \left(\frac{\partial x^2}{\partial x'^1}\right)^2 \cdot 1 \\
&= \cos^2\theta + \sin^2\theta = 1\,, \\
g'_{12}(x') &= \left(\frac{\partial x^1}{\partial x'^1}\right)\left(\frac{\partial x^1}{\partial x'^2}\right) \cdot 1 + \left(\frac{\partial x^2}{\partial x'^1}\right)\left(\frac{\partial x^2}{\partial x'^2}\right) \cdot 1 \\
&= -r\sin\theta\cos\theta + r\sin\theta\cos\theta = 0\,, \\
g'_{22}(x') &= \left(\frac{\partial x^1}{\partial x'^2}\right)^2 \cdot 1 + \left(\frac{\partial x^2}{\partial x'^2}\right)^2 \cdot 1 \\
&= r^2\sin^2\theta + r^2\cos^2\theta = r^2\,.
\end{aligned}\right\} \tag{3.40}$$

Thus, the metric tensor in plane polar coordinates is

$$g'_{ab}(r,\theta) = \begin{pmatrix} 1 & 0 \\ 0 & r^2 \end{pmatrix}\,. \tag{3.41}$$

Notice that this metric tensor correctly converts the contravariant components to covariant components. Correctly speaking, of course, the metric tensor should be written as $(1 \quad 0 \ \vdots \ 0 \quad r^2)$, but it becomes quite inconvenient to use that notation for higher dimensions, or to write the inverse matrix, which here becomes

$$g'^{ab}(r,\theta) = \begin{pmatrix} 1 & 0 \\ 0 & 1/r^2 \end{pmatrix} . \tag{3.42}$$

To check that the scalar product of the vector **A** with itself remains invariant, i.e., the length is the same whether it is expressed in Cartesian or polar coordinates, we use Eqs. (3.38) and (3.41), or equivalently Eqs. (3.38) and (3.39), to give

$$\begin{aligned}(A)^2 &= g'_{ab}(x')A'^a(x')A'^b(x') \\ &= (A'^1)^2 + r^2(A'^2)^2 \\ &= (A^x \cos\theta + A^y \sin\theta)^2 + r^2(A^y \cos\theta - A^x \sin\theta)^2/r^2 \\ &= (A^x)^2 + (A^y)^2 = g_{ij}(x)A^i(x)A^j(x) .\end{aligned} \tag{3.43}$$

This fact is shown diagramatically in Fig. 16 for two dimensions. It holds generally for all dimensions, of course.

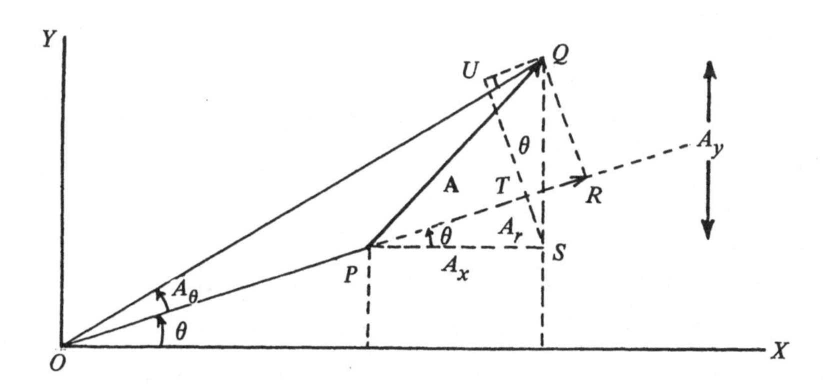


Fig. 16. The polar components are $A_r = PR$ and $A_\theta = \angle POQ$. Now, $PR = PT + TR = PS\cos\theta + UQ = A_x\cos\theta + QS\sin\theta = A_x\cos\theta + A_y\sin\theta$. $A_y\cos\theta - A_x\sin\theta = US - TS = UT$. If $A_r \ll r$, $UT/r \approx A_\theta$ radians.

Now let us consider the transformation of the metric tensor for a flat three dimensional Euclidean space into spherical polar coordinates. Here we have

$$\begin{aligned} x^1 &= x = r \sin\theta \cos\phi = x'^1 \sin x'^2 \cos x'^3 \ , \\ x^2 &= y = r \sin\theta \sin\phi = x'^1 \sin x'^2 \sin x'^3 \ , \\ x^3 &= z = r \cos\theta = x'^1 \cos x'^2 \ . \end{aligned} \tag{3.44}$$

Also, we have in Cartesian coordinates

$$g_{11} = g_{22} = g_{33} = 1, \quad g_{ij} = 0 \text{ otherwise.} \tag{3.45}$$

Thus we obtain

$$\begin{aligned} \frac{\partial x^i}{\partial x'^a} &= \begin{pmatrix} \partial x/\partial r & \partial x/\partial \theta & \partial x/\partial \phi \\ \partial y/\partial r & \partial y/\partial \theta & \partial y/\partial \phi \\ \partial z/\partial r & \partial z/\partial \theta & \partial z/\partial \phi \end{pmatrix} \\ &= \begin{pmatrix} \sin\theta \cos\phi & r\cos\theta\cos\phi & -r\sin\theta\sin\phi \\ \sin\theta\sin\phi & r\cos\theta\sin\phi & r\sin\theta\cos\phi \\ \cos\theta & -r\sin\theta & 0 \end{pmatrix} , \end{aligned} \tag{3.46}$$

which yields

$$g'_{ab}(x') = \begin{pmatrix} 1 & 0 & 0 \\ 0 & r^2 & 0 \\ 0 & 0 & r^2 \sin^2\theta \end{pmatrix} , \tag{3.47}$$

since we have

$$\left.\begin{aligned} g'_{11} &= \sin^2\theta\cos^2\phi + \sin^2\theta\sin^2\phi + \cos^2\theta = 1, \\ g'_{12} &= r(\sin\theta\cos\theta\cos^2\phi + \sin\theta\cos\theta\sin^2\phi - \sin\theta\cos\theta) = 0, \\ g'_{13} &= r\sin^2\theta(-\sin\phi\cos\phi + \sin\phi\cos\phi) = 0, \\ g'_{22} &= r^2(\cos^2\theta\cos^2\phi + \cos^2\theta\sin^2\phi + \sin^2\theta) = r^2, \\ g'_{23} &= r^2(-\sin\theta\cos\theta\sin\phi\cos\phi + \sin\theta\cos\theta\sin\phi\cos\phi) = 0, \\ g'_{33} &= r^2(\sin^2\theta\sin^2\phi + \sin^2\theta\cos^2\phi) = r^2\sin^2\theta. \end{aligned}\right\} \tag{3.48}$$

and $g'_{ab}$ is symmetric.

## Exercise 3

1. Given a 3-dimensional vector with Cartesian components $A^x$, $A^y$ and $A^z$, work out the components in a frame obtained by a rotation through $\theta_1$ in the XY-plane, then a rotation through $\theta_2$ in the YZ-plane and then a rotation through $\theta_3$ in the ZX-plane. Is the result the same if the order of rotation is reversed?

2. Prove that the Kronecker delta is unaltered by coordinate transformations.

3. Prove that the position vector, in plane polar coordinates has no polar component. Explain this fact diagramatically.

4. Given the coordinate transformation

$$p = xy, \quad q = x/y \ ,$$

   determine the metric tensor in the $(p, q)$ coordinates if $x$ and $y$ are the usual Cartesian coordinates.

5. Work out the metric of a four-dimensional Euclidean space in hyperspherical coordinates, i.e., having one radial coordinate and three angular coordinates.

6. By restricting the three-dimensional metric tensor in spherical polar coordinates to the surface of a sphere, $r = a$, determine the metric tensor for the surface of a sphere. (Remember that the resultant space is two-dimensional and not three-dimensional.)

# Chapter 4

# THE FOUR-VECTOR FORMULATION OF SPECIAL RELATIVITY

## 1. The Four-Vector Formalism

Soon after Einstein presented the Special Theory of Relativity, his Mathematics teacher, Minkowski, published a reformulation of the theory in more mathematical terms using 4-vectors. Einstein's initial reaction was to reject this approach. However, he later used it extensively, particularly for the formulation of the General Theory of Relativity. The reason for Einstein's initial resistance was his lack of familiarity with geometry. We have developed nearly all the aspects of geometry required for the formulation of the Special Theory. Only one more point needs to be developed, which we shall proceed with here.

In dealing with the theory of surfaces it is found that two types of metrics arise (corresponding to the two fundamental forms). One is *positive definite*, i.e., for every vector a the metric gives

$$(a)^2 \geq 0 \text{ and } (a)^2 = 0 \Longleftrightarrow \mathrm{a} = 0 \ . \tag{4.1}$$

The other type of metric is *indefinite*, i.e., Eq. (4.1) does not hold at all. In fact neither part holds in general. For these metrics points are said to be elliptic, parabolic or hyperbolic according as for any $\mathrm{a} \neq 0$, $(a)^2 > 0$, $(a)^2 \geq 0$ or $(a)^2 \geq 0$. A space in which all points are elliptic, or parabolic, or hyperbolic, is called an elliptic, parabolic or hyperbolic space. We will not really need to deal with virtually any part of the geometry discussed here, except in passing. However, familiarity with the terminology given will be useful. Also, the further developments in Relativity, particularly of General Relativity, require the use of extensions of the geometry presented here. We now return to Special Relativity.

Consider the product of Eqs. (2.5) and (2.6)

$$c^2t'^2 - x'^2 = \lambda\mu(c^2t^2 - x^2) \ . \tag{4.2}$$

Also, from Eqs. (2.8) we have

$$a^2 - b^2 = (\lambda+\mu)^2/4 - (\lambda-\mu)^2/4 = \lambda\mu \ . \tag{4.3}$$

Further, from Eqs. (2.12) and (2.20) we see that

$$a^2 - b^2 = a^2(1 - v^2/c^2) = 1 \ . \tag{4.4}$$

Thus, using Eqs. (4.3) and (4.4) in Eq. (4.2) we obtain

$$c^2t'^2 - x'^2 = c^2t^2 - x^2 \ . \tag{4.5}$$

Had we dealt with a 3-dimensional vector, $\mathbf{x}$, instead of only the $x$-coordinate, clearly $x^2$ would be replaced by

$$\mathbf{x}\cdot\mathbf{x} = x^2 + y^2 + z^2 \ . \tag{4.6}$$

Thus, we would have had, instead of Eq. (4.5),

$$c^2t'^2 - x'^2 - y'^2 - z'^2 = c^2t^2 - x^2 - y^2 - z^2 \ . \tag{4.7}$$

We would then say that the invariant quantity corresponding to the length could be given for the *4-vector*, $x^\mu$,

$$x^\mu = (x^0, x^1, x^2, x^3) = (ct, x, y, z) \ , \tag{4.8}$$

with a $g_{\mu\nu}$ defined by

$$g_{00} = 1 = -g_{11} = -g_{22} = -g_{33}, \ g_{\mu\nu} = 0 \text{ if } \mu \neq \nu \ , \tag{4.9}$$

so that the metric is hyperbolic. Now

$$g_{\mu\nu} x^\mu x^\nu = g_{\mu\nu} x'^\mu x'^\nu \quad . \tag{4.10}$$

In other words this quantity is invariant under Lorentz transformations. Notice that since $g_{ij} \neq 1$ for $i, j = 1, 2, 3$, there is a difference of sign between the spatial parts of covariant and contravariant vectors.

## 2. The Lorentz Transformations in 4-Vectors

We now need to describe the Lorentz transformations in terms of 4-vectors. We can write

$$x'^\mu = \Lambda^\mu_\alpha \, x^\alpha \quad , \tag{4.11}$$

i.e., if $x^\alpha$ is a column vector $\Lambda^\mu_\alpha$ is a square matrix. Let us look at the form of $\Lambda^\mu_\alpha$ in the case that the motion is purely in the $x$-direction. For convenience we write $\beta = v/c$ instead of $v$ and so $\gamma = (1-\beta^2)^{-\frac{1}{2}}$. Thus, we are representing the speed as a fraction of light speed. This is more convenient than the previous way since $\beta$ is always less than one and has no units of measurement. Similarly, it is convenient to have $t$ measured in the same units as $x$, $y$ and $z$. Thus, we write $T = ct$, which is the distance covered by light in time $t$, e.g. a light second is $3 \times 10^5$ km (or $3 \times 10^8$ m or $3 \times 10^{10}$ cm) approximately. The Lorentz transformations now become

$$\begin{pmatrix} T' \\ x' \\ y' \\ z' \end{pmatrix} = \begin{pmatrix} \gamma & -\beta\gamma & 0 & 0 \\ -\beta\gamma & \gamma & 0 & 0 \\ 0 & 0 & 1 & 0 \\ 0 & 0 & 0 & 1 \end{pmatrix} \begin{pmatrix} T \\ x \\ y \\ z \end{pmatrix} \quad . \tag{4.12}$$

Notice that the matrix is in block diagonal form in Eq. (4.12), with the lower block being the identity. Thus, the only relevant part is in the upper block. For the present let us disregard the lower block, i.e., let us consider only one spatial dimension. If we rewrite the matrix equation, Eq. (4.12), we have

$$\begin{pmatrix} T' \\ x' \end{pmatrix} = \begin{pmatrix} \gamma & -\beta\gamma \\ -\beta\gamma & \gamma \end{pmatrix} \begin{pmatrix} T \\ x \end{pmatrix} \quad . \tag{4.13}$$

This matrix equation may be inverted easily. The determinant of the Lorentz transformation is

$$\det\begin{pmatrix} \gamma & -\beta\gamma \\ -\beta\gamma & \gamma \end{pmatrix} = \gamma^2 - \beta^2\gamma^2 = \gamma^2(1-\beta^2) = 1 \ . \qquad (4.14)$$

Thus we see that

$$\begin{pmatrix} T \\ x \end{pmatrix} = \begin{pmatrix} \gamma & \beta\gamma \\ \beta\gamma & \gamma \end{pmatrix} \begin{pmatrix} T' \\ x' \end{pmatrix} \ . \qquad (4.15)$$

In other words, to invert the Lorentz transformations for motion in one spatial dimension we need merely invert the sign of $\beta$, or equivalently of $v$. This is an important check of consistency, since it is required by the special relativity principle. After all, we could have chosen to regard $O'$ as being at rest and $O$ as being in uniform motion.

It is interesting to observe that the hyperbolic functions cosh and sinh obey the same relation as $\gamma$ and $\beta\gamma$, i.e.,

$$\cosh^2\theta - \sinh^2\theta = 1 \ . \qquad (4.16)$$

Thus, we could replace $\gamma$ and $\beta$ in Eq. (4.15) by $\cosh\theta$ and $\sinh\theta$ and write

$$\begin{pmatrix} T' \\ x' \end{pmatrix} = \begin{pmatrix} \cosh\theta & -\sinh\theta \\ -\sinh\theta & \cosh\theta \end{pmatrix} \begin{pmatrix} T \\ x \end{pmatrix} \ . \qquad (4.17)$$

Noticing that $\cosh\theta = \cos i\theta$ and $\sinh\theta = -i\sin i\theta$, we see that the Lorentz transformations correspond to a rotation through an imaginary angle, just like we have for ordinary rotations

$$\begin{pmatrix} x' \\ y' \end{pmatrix} = \begin{pmatrix} \cos\theta & \sin\theta \\ -\sin\theta & \cos\theta \end{pmatrix} \begin{pmatrix} x \\ y \end{pmatrix} \ . \qquad (4.18)$$

Thus, we should not think of space and time as separate, but rather of space-time jointly. The essence of the argument is that they can be 'mixed together' through the physical process of motion in a straight line just as the spatial coordinates can be 'mixed together' by the physical process of rotation.

Since the Lorentz transformations being considered can be represented by square matrices with unit determinant, they form a group in which the inverse matrix corresponds to the physical inverse Lorentz transformations. Here the identity transformation is the case when $\beta = 0$ (or $v = 0$, or $\theta = 0$). Associativity is anyhow guaranteed. However, we do need to check that the combination of two Lorentz transformations is, in fact, a Lorentz transformation. Let the two speeds be $v_1$ and $v_2$ giving $\beta_1$ and $\beta_2$ and thus $\theta_1$ and $\theta_2$. The combination is achieved by multiplication. Then

$$\begin{aligned}
&\begin{pmatrix} \cosh\theta_1 & -\sinh\theta_1 \\ -\sinh\theta_1 & \cosh\theta_1 \end{pmatrix} \begin{pmatrix} \cosh\theta_2 & -\sinh\theta_2 \\ -\sinh\theta_2 & \cosh\theta_2 \end{pmatrix} \\
&= \begin{pmatrix} (\cosh\theta_1\cosh\theta_2 + \sinh\theta_1\sinh\theta_2) & -(\sinh\theta_1\cosh\theta_2 + \cosh\theta_1\sinh\theta_2) \\ -(\sinh\theta_1\cosh\theta_2 + \cosh\theta_1\sinh\theta_2) & (\cosh\theta_1\cosh\theta_2 + \sinh\theta_1\sinh\theta_2) \end{pmatrix} \\
&= \begin{pmatrix} \cosh(\theta_1+\theta_2) & -\sinh(\theta_1+\theta_2) \\ -\sinh(\theta_1+\theta_2) & \cosh(\theta_1+\theta_2) \end{pmatrix} .
\end{aligned} \tag{4.19}$$

Thus we do, in fact, get the required property with the resultant being the sum of $\theta_1$ and $\theta_2$, exactly as for rotation. Now

$$\beta = \beta\gamma/\gamma = \sinh\,\theta/\cosh\,\theta = \tanh\,\theta\ . \tag{4.20}$$

Thus, we have

$$\beta = \tanh\,\theta = \tanh(\theta_1+\theta_2) = \frac{\tanh\,\theta_1 + \tanh\,\theta_2}{1+\tanh\,\theta_1\,\tanh\,\theta_2}\ . \tag{4.21}$$

If we write this in terms of $v_1$, $v_2$ and $v$

$$v = \frac{\beta_1+\beta_2}{1+\beta_1\beta_2}c = \frac{v_1+v_2}{1+v_1v_2/c^2}\ , \tag{4.22}$$

which gives the velocity addition formula, Eq. (2.38). The fact that the Lorentz transformations form a group which yields the above velocity addition formula was noted independently (and somewhat earlier) by Poincaré. Poincaré also noticed that Eq. (4.22) implies that $c$ will be the same for all observers.

Let us return to the full Lorentz transformations. Here we must write $\boldsymbol{\beta} = \mathbf{v}/c$. Then we can rewrite Eqs. (2.48) and (2.49) in terms of the separate components $T$, $x^1$, $x^2$, $x^3$ and $\beta^1$, $\beta^2$, $\beta^3$. We retain $\gamma$ as it was. In this notation, Eqs. (2.48) and (2.49) become

$$\left.\begin{aligned}
T' &= \gamma T - \beta^1\gamma x^1 - \beta^2\gamma x^2 - \beta^3\gamma x^3 \ , \\
x'^1 &= -\beta^1\gamma T + \left\{1 + (\gamma-1)\frac{(\beta^1)^2}{\boldsymbol{\beta}\cdot\boldsymbol{\beta}}\right\}x^1 + (\gamma-1)\frac{\beta^1\beta^2}{\boldsymbol{\beta}\cdot\boldsymbol{\beta}}x^2 \\
&\quad + (\gamma-1)\frac{\beta^1\beta^3}{\boldsymbol{\beta}\cdot\boldsymbol{\beta}}x^3 \ , \\
x'^2 &= -\beta^2\gamma T + (\gamma-1)\frac{\beta^1\beta^2}{\boldsymbol{\beta}\cdot\boldsymbol{\beta}}x^1 + \left\{1 + (\gamma-1)\frac{(\beta^2)^2}{\boldsymbol{\beta}\cdot\boldsymbol{\beta}}\right\}x^2 \\
&\quad + (\gamma-1)\frac{\beta^2\beta^3}{\boldsymbol{\beta}\cdot\boldsymbol{\beta}}x^3 \ , \\
x'^3 &= -\beta^3\gamma T + (\gamma-1)\frac{\beta^1\beta^3}{\boldsymbol{\beta}\cdot\boldsymbol{\beta}}x^1 + (\gamma-1)\frac{\beta^2\beta^3}{\boldsymbol{\beta}\cdot\boldsymbol{\beta}}x^2 \\
&\quad + \left\{1 + (\gamma-1)\frac{(\beta^3)^2}{\boldsymbol{\beta}\cdot\boldsymbol{\beta}}\right\}x^3 \ .
\end{aligned}\right\} \tag{4.23}$$

In matrix form Eqs. (4.23) can be written as

$$\begin{pmatrix} T' \\ x'^1 \\ x'^2 \\ x'^3 \end{pmatrix} =$$

$$\begin{pmatrix}
\gamma & -\beta^1\gamma & -\beta^2\gamma & -\beta^3\gamma \\
-\beta^1\gamma & 1 + \frac{(\beta^1)^2(\gamma-1)}{\boldsymbol{\beta}\cdot\boldsymbol{\beta}} & \frac{\beta^1\beta^2(\gamma-1)}{\boldsymbol{\beta}\cdot\boldsymbol{\beta}} & \frac{\beta^1\beta^3(\gamma-1)}{\boldsymbol{\beta}\cdot\boldsymbol{\beta}} \\
-\beta^2\gamma & \frac{\beta^1\beta^2(\gamma-1)}{\boldsymbol{\beta}\cdot\boldsymbol{\beta}} & 1 + \frac{(\beta^2)^2(\gamma-1)}{\boldsymbol{\beta}\cdot\boldsymbol{\beta}} & \frac{\beta^2\beta^3(\gamma-1)}{\boldsymbol{\beta}\cdot\boldsymbol{\beta}} \\
-\beta^3\gamma & \frac{\beta^1\beta^3(\gamma-1)}{\boldsymbol{\beta}\cdot\boldsymbol{\beta}} & \frac{\beta^2\beta^3(\gamma-1)}{\boldsymbol{\beta}\cdot\boldsymbol{\beta}} & 1 + \frac{(\beta^3)^2(\gamma-1)}{\boldsymbol{\beta}\cdot\boldsymbol{\beta}}
\end{pmatrix}
\begin{pmatrix} T \\ x^1 \\ x^2 \\ x^3 \end{pmatrix} \tag{4.24}$$

which gives the explicit expression for the matrix $\Lambda^\mu_\nu$. It can be checked that these transformations have a unit determinant and the inverse is given by replacing $\boldsymbol{\beta}$ by $-\boldsymbol{\beta}$. The easiest way to do so is, just to multiply the two matrices

$$\begin{pmatrix} \gamma & \beta^1\gamma & \beta^2\gamma & \beta^3\gamma \\ \beta^1\gamma & \left\{1+\frac{(\beta^1)^2(\gamma-1)}{\boldsymbol{\beta}\cdot\boldsymbol{\beta}}\right\} & \frac{\beta^1\beta^2(\gamma-1)}{\boldsymbol{\beta}\cdot\boldsymbol{\beta}} & \frac{\beta^1\beta^3(\gamma-1)}{\boldsymbol{\beta}\cdot\boldsymbol{\beta}} \\ \beta^2\gamma & \frac{\beta^1\beta^2(\gamma-1)}{\boldsymbol{\beta}\cdot\boldsymbol{\beta}} & \left\{1+\frac{(\beta^2)^2(\gamma-1)}{\boldsymbol{\beta}\cdot\boldsymbol{\beta}}\right\} & \frac{\beta^2\beta^3(\gamma-1)}{\boldsymbol{\beta}\cdot\boldsymbol{\beta}} \\ \beta^3\gamma & \frac{\beta^1\beta^3(\gamma-1)}{\boldsymbol{\beta}\cdot\boldsymbol{\beta}} & \frac{\beta^2\beta^3(\gamma-1)}{\boldsymbol{\beta}\cdot\boldsymbol{\beta}} & \left\{1+\frac{(\beta^3)^2(\gamma-1)}{\boldsymbol{\beta}\cdot\boldsymbol{\beta}}\right\} \end{pmatrix}$$

$$\times \begin{pmatrix} \gamma & -\beta^1\gamma & -\beta^2\gamma & -\beta^3\gamma \\ -\beta^1\gamma & \left\{1+\frac{(\beta^1)^2(\gamma-1)}{\boldsymbol{\beta}\cdot\boldsymbol{\beta}}\right\} & \frac{\beta^1\beta^2(\gamma-1)}{\boldsymbol{\beta}\cdot\boldsymbol{\beta}} & \frac{\beta^1\beta^3(\gamma-1)}{\boldsymbol{\beta}\cdot\boldsymbol{\beta}} \\ -\beta^2\gamma & \frac{\beta^1\beta^2(\gamma-1)}{\boldsymbol{\beta}\cdot\boldsymbol{\beta}} & \left\{1+\frac{(\beta^2)^2(\gamma-1)}{\boldsymbol{\beta}\cdot\boldsymbol{\beta}}\right\} & \frac{\beta^2\beta^3(\gamma-1)}{\boldsymbol{\beta}\cdot\boldsymbol{\beta}} \\ -\beta^3\gamma & \frac{\beta^1\beta^3(\gamma-1)}{\boldsymbol{\beta}\cdot\boldsymbol{\beta}} & \frac{\beta^2\beta^3(\gamma-1)}{\boldsymbol{\beta}\cdot\boldsymbol{\beta}} & \left\{1+\frac{(\beta^3)^2(\gamma-1)}{\boldsymbol{\beta}\cdot\boldsymbol{\beta}}\right\} \end{pmatrix}$$

$$= \begin{pmatrix} 1 & 0 & 0 & 0 \\ 0 & 1 & 0 & 0 \\ 0 & 0 & 1 & 0 \\ 0 & 0 & 0 & 1 \end{pmatrix} . \tag{4.25}$$

## 3. The Lorentz and Poincaré Groups

As was shown, for every Lorentz transformation, $\Lambda^\mu_\nu$, there exists an inverse Lorentz transformation $(\Lambda^\mu_\nu)^{-1}$, obtained by changing the three parameters $\beta^i$ $(i = 1, 2, 3)$ to $-\beta^i$, in $\Lambda^\mu_\nu$. Formally we have

$$\Lambda^\mu_\nu (\Lambda^{-1})^\nu_\rho = \delta^\mu_\rho \ , \tag{4.26}$$

where $(\Lambda^{-1})^\nu_\rho = (\Lambda^\rho_\nu)^{-1}$ and $\delta^\mu_\rho$ is the Kronecker delta. Since the Lorentz transformations are expressible as matrices they are closed under multiplication and are associative. Also, the identity element $(\delta^\mu_\rho)$ exists and inverses exist. Thus, they form a group. In fact this set of matrices has a unit determinant. Also, we could include rotations between the $X$-, $Y$- and $Z$-axes. Thus, we would generally have, for all the Lorentz transformations, three parameters for the rotations between $T$- and $X$-, $T$- and $Y$-, and $T$- and $Z$-axes and three more for the purely spatial rotations. The group of all Lorentz transformations of this type therefore has six parameters. In any position 4-vector we can change $T$ to $-T$ and obtain, from a future-directed vector, a past-directed vector and vice-versa. Since these transformations are *not* obtainable from the previously discussed transformations, they must be included separately. They clearly also leave $g_{\mu\nu}\, x^\mu x^\nu$ invariant, but their determinant is $-1$ (since $\mathbf{x} \rightarrow \mathbf{x}$). This is called a time reflection. Two reflections give an identity. Clearly, these transformations also form a group. (Notice that $\mathbf{x} \rightarrow -\mathbf{x}$ can be obtained from the rotation matrix by taking the values of the parameters corresponding to a total rotation through $\pi$.) The group of all transformations that leaves $g_{\mu\nu}\, x^\mu x^\nu$ invariant is called the *Lorentz-group*. The reflections give the *improper Lorentz group* while the rotations give the *proper Lorentz group*.

We could also consider the set of all translations in the 4-dimensional space-time. These would change the position vector by addition with the translating vector $a^\mu$,

$$x^\mu \rightarrow x'^\mu = x^\mu + a^\mu \ . \tag{4.27}$$

However, a physical vector would be left invariant under such a transformation, i.e. if $A^\mu$ connects a point in the space-time, $P$, to another point $Q$, $A'^\mu = A^\mu$. This is easily seen if we take the position vector of $P$ to be $x^\mu$ and of $Q$ to be $y^\mu$

$$A^\mu = y^\mu - x^\mu \ . \tag{4.28}$$

Using Eq. (4.27) to transform $x^\mu$ and $y^\mu$ we see that

$$\begin{aligned} A'^\mu &= y'^\mu - x'^\mu = (y^\mu + a^\mu) - (x^\mu + a^\mu) \\ &= y^\mu - x^\mu = A^\mu \ . \end{aligned} \tag{4.29}$$

According to the theory of Special Relativity all physical laws are invariant under Lorentz transformations, since these are transformations from one inertial frame to another. According to the basic assumptions of physics all physical laws are also invariant under spatial and temporal translations. These translations form a group, which is the group of 4-vectors under addition. The identity transformation is the zero vector and every translation, $a^\mu$ has an inverse, $-a^\mu$. The group of translations for 4-vectors depends on four parameters.

The entire group under which physical laws are invariant according to Special Relativity is the ten parameter *Poincaré group*, presented in 1905 by Poincaré,

$$P(10) = L(6) \otimes T(4) \ , \tag{4.30}$$

where the numbers inside the brackets refer to the number of independent parameters of the group.

The improper Lorentz group does not correspond to a continuous transformation like rotations do. Thus it is a discrete group. The $L(6)$ referred to in Eq. (4.30) is therefore only the proper Lorentz group. It is the group of all rotations in Minkowski space, i.e., rotations which leave invariant the form given by Eq. (4.7) which has the first square positive and the other three negative. The group of rotations in $n$-dimensional Euclidean space is called the *orthogonal group*, SO($n$), where the 'S' signifies that the group is unimodular, i.e. the matrix representing an element of this group has a unit determinant. In Minkowski space the group is SO(1,3), the '1' referring to the time dimension and the '3' to the space dimensions. Thus, we should really use this group rather than the entire $L(6)$. Now, we saw that SO(1,3) contains, as the relevant subgroups, the group of rotations among $x, y, z$ and also the rotations of time with each of the $x, y, z$. Thus

$$\mathrm{SO}(1,3) \supseteq \mathrm{SO}(3) \otimes \mathrm{SO}(3) \ , \tag{4.31}$$

where one SO(3) refers to space and the other to space-time. The semidirect product in Eq. (4.30), written as '$\otimes$', indicates that the elements of SO(1,3) do not commute with $T(4)$. This may be seen by considering $x'^\mu$ defined by

$$x'^\mu = \Lambda^\mu_\nu x^\nu \ , \tag{4.32}$$

and then

$$x''^\mu = x'^\mu + a^\mu = \Lambda^\mu_\nu x^\nu + a^\mu \ . \tag{4.33}$$

On the other hand if we define

$$\hat{x}^{\mu} = x^{\mu} + a^{\mu} \ , \tag{4.34}$$

and then define

$$\hat{\hat{x}}^{\mu} = \Lambda^{\mu}_{\nu}\, \hat{x}^{\nu} = \Lambda^{\mu}_{\nu}\, x^{\nu} + \Lambda^{\mu}_{\nu}\, a^{\nu} \ , \tag{4.35}$$

clearly we will have, in general,

$$\hat{\hat{x}}^{\mu} \neq x''^{\mu} \ . \tag{4.36}$$

If we had a direct product, as for example in Eq. (4.31), the two would have to be equal.

Now, according to Noether's theorem in classical mechanics, for every parameter of symmetry, i.e., for every generator of the symmetry group, there is an invariant, or conserved, quantity. For the translations in space-time we have the conservation of energy-momentum. For the spatial SO(3) we have angular momentum conservation, while for the other SO(3) we get spin conservation. Since SO(1,3) has elements which do not belong to either SO(3) alone, we can have spin and angular momentum 'mixing', i.e., each separately may not be conserved, but the total is conserved. Thus, when relativistic quantum mechanics is considered, we get the electron spin predicted and spin-orbit coupling in atomic (and nuclear) spectra. We will not discuss further, the physical implications of invariance under the Poincaré group, but will proceed on to the geometrical structure of the Special Relativistic, Minkowski, space-time.

## 4. The Null Cone Structure

Consider, first, an infinitesimal vector, $dx^{\mu}$, which is invariant under the Lorentz transformations, in the sense that its square magnitude remains invariant, but not its components,

$$ds^2 = g_{\mu\nu}\, dx^{\mu}\, dx^{\nu} = g_{\mu\nu}\, dx'^{\mu}\, dx'^{\nu} \ , \tag{4.37}$$

and is invariant under translations in the sense that

$$dx^\mu \to dx'^\mu = dx^\mu \ . \tag{4.38}$$

Thus $ds^2$ is invariant under the full Poincaré group.

A 4-vector, $dx^\mu$ is said to be *time-like, null(light-like)* or *space-like* according as $ds^2 \gtreqless 0$. Let us see what this means geometrically. For convenience we shall write

$$dx^\mu = (dx^0, dx^i) = (cdt, d\mathbf{x}) \ . \tag{4.39}$$

Now, we see that we have for

$$\left.\begin{array}{ll} \text{time-like vectors} & c^2dt^2 - d\mathbf{x}\cdot d\mathbf{x} > 0 \\ \text{null vectors} & c^2dt^2 - d\mathbf{x}\cdot d\mathbf{x} = 0 \\ \text{space-like vectors} & c^2dt^2 - d\mathbf{x}\cdot d\mathbf{x} < 0 \end{array}\right\} \ . \tag{4.40}$$

Dividing through by $dt^2$ and transposing the second terms, we get for

$$\left.\begin{array}{ll} \text{time-like vectors} & c^2 > (d\mathbf{x}/dt)\cdot(d\mathbf{x}/dt) \\ \text{null vectors} & c^2 = (d\mathbf{x}/dt)\cdot(d\mathbf{x}/dt) \\ \text{space-like vectors} & c^2 < (d\mathbf{x}/dt)\cdot(d\mathbf{x}/dt) \end{array}\right\} \ . \tag{4.41}$$

Now, $d\mathbf{x}/dt$ corresponds to a velocity, $\mathbf{v}$. For time-like vectors the magnitude of $\mathbf{v}$ is less than $c$. Thus $dx^\mu$ can represent the actual path of a physical object in space over time. For null vectors the magnitude of $\mathbf{v}$ is equal to $c$. Thus $dx^\mu$ can represent the path of a physical object travelling at light-speed. In the case of null vectors, it will be seen that the classical concept of a particle will not be applicable. However, we can still have energy going at light speed, even classically. For space-like vectors the magnitude of $\mathbf{v}$ is *greater than* $c$! This is not possible for a physical object such as a particle. Nevertheless, it could describe an object of spatial extension $d\mathbf{x}$ seen from a frame such that the two ends are not seen simultaneously, but with an interval $dt$.

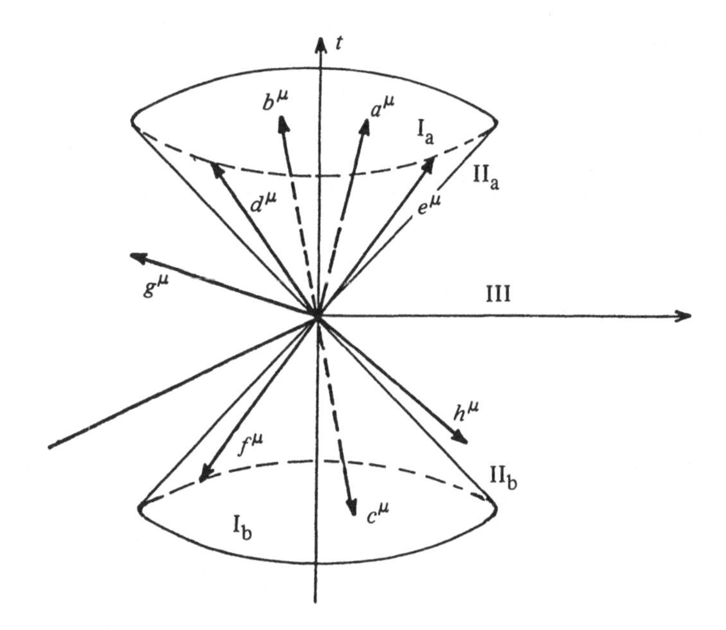


Fig. 17. The null cone at O. The space-time is divided into the three regions, time-like (I), null (II) and space-like (III). Region I is 4-dimensional and is the interior of the null cone. It is further divided into two regions, the future-directed (Ia) and the past-directed (Ib). The vector $b^\mu$ is obtained from $a^\mu$ by a proper Lorentz transformation, while $c^\mu$ is obtained from $a^\mu$ by an improper Lorentz transformation. Region II is the 3-dimensional surface of the null cone, again divided into future directed (IIa) and past-directed (IIb). Vectors $d^\mu$ and $e^\mu$ are related by proper Lorentz transformations. Region III is connected and all vectors like $g^\mu$ and $h^\mu$ are related by Lorentz transformations (and a scaling). Vectors in regions I, II and III are called time-like, null and space-like, respectively.

Since Lorentz transformations leave $ds^2$ invariant, the type of vector is not changed by a Lorentz transformation. Also, since translations leave $ds^2$ invariant we can always translate the origin in space-time to the origin of $dx^\mu$. In the frame of reference with its origin at the point under consideration, measuring distance in light units, if we plot $t$ against $x$, null vectors will lie along the lines $t = \pm x$. Time-like vectors will lie above those lines and space-like vectors below them. Actually we should plot $t$ against $\pm\mathbf{x}$. For easier visualisation we suppress one of the three coordinates and use only two axes for $\mathbf{x}$. In that case we get a cone as shown in Fig. 17. Time-like vectors lie inside the cone and space-like vectors outside. Null vectors lie on the surface of the cone. Notice that Lorentz transformations must leave this structure, called the ***null cone structure***, invariant. They rotate vectors inside the cone into other vectors inside the cone, those outside are rotated to others outside, while those on the surface will be rotated on the surface. It is clear that no Lorentz transformation of the rotation type can then convert a vector in (on) the upper cone to one in (on) the lower cone. Thus, there are two types of time-like (null) vectors: those that are future-directed (i.e., have $dx^0 > 0$); and those that are past-directed (i.e., have $dx^0 < 0$). [Clearly, if $dx^0 = 0$, $ds^2 \leq 0$ and $ds^2 = 0 \Longleftrightarrow d\mathbf{x} = 0$.] There are, thus, three disjoint regions which contain the vectors on which the representations of the Lorentz group act. Two of them are disjoint unions (of two regions each) for the proper Lorentz group. The improper Lorentz group, however, connects those regions.

## 5. The Search For Absolutes — Proper-Time

Consider the invariant interval, $ds^2$, in the rest-frame of the observer $O$. There $dx^0 = cdt$ and $dx^i = 0$. Inserting these into Eq. (4.37) we see that in the frame of $O$

$$ds^2 = c^2 dt^2 \ . \tag{4.42}$$

In the frame of $O'$, we have, of course,

$$ds^2 = c^2 dt'^2 - dx'^2 - dy'^2 - dz'^2 \ . \tag{4.43}$$

We define the ***proper-time***, which is an invariant quantity, as

$$d\tau^2 = ds^2/c^2 \ . \tag{4.44}$$

Thus, comparing Eqs. (4.42) and (4.44), we see that the proper-time is exactly the time measured in the rest-frame. ***This is an absolute (invariant) quantity!*** Any other observer can measure the coordinate time $dt'$ and the spatial displacement $d\mathbf{x}'$ and obtain the absolute quantity through Eqs. (4.37) and (4.44).

The question is often raised whether time-dilation etc. are real or apparent effects. The question is rephrased, also, in the form of whether all quantities are relative or some are absolute. It is necessary to distinguish between the two questions. There are definitely some absolute quantities. Any 4-vector is an absolute quantity though its components appear to change. The language used to deal with such absolute quantities, in general, is Differential Geometry. To deal with general motion and not only uniform linear motion, we will need to use that language. However, for the purpose of Special Relativity, we need not bother with that language. As regards the earlier question, it is more complicated. The problem lies in the meaning assigned to the terms 'real' and 'apparent'. The effects are certainly not *dynamical* effects. They are entirely *kinematic* in origin. By this it is meant that there are *no forces* responsible, but merely the change of frame of the observer. However, this fact does not make the effect less *physical*. The point is that *any measurement* made by the moving observer will show the relativistic effects. There is no operational significance to the non-invariant quantities in other frames. Speaking very loosely, the absolute quantities can be regarded as 'most real' while all other quantities retain their 'reality' for the appropriate observer. In this sense, the theory of relativity recognizes that some quantities that were classically taken to be absolute *a priori* are in fact relative, and then provides the statements that are observer-independent, i.e., absolute.

The procedure will now be to use the absolute quantities to derive physical statements that enable us to obtain answers to questions of mechanics. We will first deal with kinematics and then with dynamics. In the kinematic section we will look at some physical consequences and then proceed on to dynamics. The same procedure of defining and using absolute quantities will be adopted for dynamics.

## Exercise 4

1. Let an observer $A$ see observer $B$ moving with a speed $u$ in the $x$-direction, observer $B$ see observer $C$ moving in the $y$-direction and observer $C$ see observer $D$ moving in the $z$-direction. Work out the Lorentz transformation matrix of the motion of $D$ relative to $A$ and of $A$ relative to $D$. Are the two matrices inverses of each other?

2. Explain the relativity of simultaneity in terms of the null cone.

3.* Explain, in terms of the null cone structure, why a massive particle can never go at the speed of light and a massless particle can never go less than the speed of light.

4. Given a contravariant vector using spherical polar coordinates

$$A^{\mu} = \begin{pmatrix} A^t \\ A^r \\ A^{\theta} \\ A^{\phi} \end{pmatrix} ,$$

work out the covariant components $A_{\mu}$ and the magnitude of the vector. Work out the value of the time component of this vector if it is a null vector.

5.* Prove that if the sum of two velocity 4-vectors is a velocity 4-vector the angle between them is $2\pi/3$, while if their difference is a velocity 4-vector the angle between them is $\pi/3$.

* These questions use some concepts explained in the next chapter.

# Chapter 5

# APPLICATIONS OF SPECIAL RELATIVITY

## 1. Relativistic Kinematics

Having obtained some invariant quantities for kinematics, we can proceed to define others. Since the 4-vector, $dx^\mu$, is such a quantity and the proper-time interval $d\tau$ is also invariant,

$$\dot{x}^\mu \equiv dx^\mu/d\tau = (cdt/d\tau, d\mathbf{x}/d\tau) \tag{5.1}$$

is also an invariant quantity. Now, using Eq. (2.24), we see that

$$dt/d\tau = \gamma \ . \tag{5.2}$$

Also, we have

$$d\mathbf{x}/d\tau = (d\mathbf{x}/dt)(dt/d\tau) = \gamma\mathbf{v} \ . \tag{5.3}$$

Thus the *4-vector velocity* is defined by

$$v^\mu = \dot{x}^\mu = \gamma(c, \mathbf{v}) \ . \tag{5.4}$$

Now, we can work out the magnitude of this vector

$$v^\mu \, v^\nu \, g_{\mu\nu} = \gamma^2(c^2 - \mathbf{v}\cdot\mathbf{v}) = c^2 \ . \tag{5.5}$$

Thus all velocity 4-vectors have a constant magnitude. Defining the velocity 4-vector $V^\mu = v^\mu/c$, we see that $V^\mu$ has unit magnitude, i.e., it is a unit 4-vector. The choice of 4-vector velocity is, then, one of the angle it makes with the time-axis in the null cone. (It corresponds to the unit tangent vector in the theory of curves.)

Now, the *momentum* of a particle is its mass, $m$, times its velocity. Thus, the *4-vector momentum* is defined by

$$p^\mu = mv^\mu = (p^0, \mathbf{p}) = (\gamma mc, \gamma m\mathbf{v}) \ . \tag{5.6}$$

Now the momentum of the moving particle, by definition, must be the mass times the velocity. But, in this case it will be the 'moving-mass', $m'$, rather than the 'rest-mass', $m$. Hence

$$\mathbf{p} = m'\mathbf{v} = \gamma m\mathbf{v} \ , \tag{5.7}$$

which gives the formula for the moving mass

$$m' = \gamma m \ . \tag{5.8}$$

The question arises what $p^0$ corresponds to. To interpret this we shall make a slight detour through classical mechanics.

Consider the Poisson bracket of any function $A(q^i, p_i)$ with the generalised position and momentum, $q^j$ and $p_j$ respectively,

$$\left.\begin{aligned} (A, q^j) &= \frac{\partial A}{\partial q^i}\frac{\partial q^j}{\partial p_i}{}^{\nearrow 0} - \frac{\partial A}{\partial p_i}\frac{\partial q^j}{\partial q^i} = -\frac{\partial A}{\partial p_j} \\ (A, p_j) &= \frac{\partial A}{\partial q^i}\frac{\partial p_j}{\partial p_i} - \frac{\partial A}{\partial p_i}\frac{\partial p_j}{\partial q^i}{}^{\nearrow 0} = \frac{\partial A}{\partial q_j} \ . \end{aligned}\right\} \tag{5.9}$$

The first term on the right side in the first equation and the second in the second equation being zero since $q^i$ and $p_i$ are independent variables. Thus, the Poisson bracket of $A$ with position corresponds to its derivative with respect to momentum and is multiplied by $-1$. Conversely, the Poisson bracket with momentum corresponds to differentiation with respect to position. Now consider the Poisson bracket with the Hamiltonian,

$$(A, H) = \frac{\partial A}{\partial q^i}\frac{\partial H}{\partial p_i} - \frac{\partial A}{\partial p_i}\frac{\partial H}{\partial q^i} \ , \tag{5.10}$$

bearing in mind the Hamilton-Jacobi equations

$$\frac{\partial H}{\partial p_i} = \dot{q}^i \; , \quad \frac{\partial H}{\partial q^i} = -\dot{p}_i \; . \tag{5.11}$$

Inserting Eq. (5.11) into Eq. (5.10), we get

$$(A, H) = \frac{\partial A}{\partial q^i}\frac{dq^i}{dt} + \frac{\partial A}{\partial p_i}\frac{dp_i}{dt} = \frac{dA}{dt} \; . \tag{5.12}$$

Thus, the Poisson bracket of $A$ with the Hamiltonian corresponds to a derivative with respect to time. In other words energy is to momentum what time is to position (since the Hamiltonian is just the total energy of the system). Clearly, $H/c$ corresponds to $T$ which is $x^0$. Now the energy corresponding to the Hamiltonian is written as $E$. Thus, we have

$$p^0 = E/c \; . \tag{5.13}$$

Putting Eq. (5.13) into Eq. (5.6), we see that

$$E = \gamma mc^2 = m'c^2 \; , \tag{5.14}$$

the famous energy-mass relation of Einstein.

We notice, from Eq. (5.14), that even at rest ($v = 0$) there will be a residual energy, called the *rest-energy*

$$E_0 = mc^2 \; . \tag{5.15}$$

Thus, the kinetic energy is

$$T = E - E_0 = (\gamma - 1)\, mc^2 \; . \tag{5.16}$$

Consider this expression for kinetic energy for small $v$, to lowest order in $v/c$. Now we have

$$\gamma - 1 = (1 - v^2/c^2)^{-\frac{1}{2}} - 1 = \frac{1}{2}v^2/c^2 + 0(v/c)^4 \; . \tag{5.17}$$

Thus we see that we obtain the classical result

$$T = \frac{1}{2}mv^2 + 0(v/c)^2 \; . \tag{5.18}$$

The relativistic correction to this result would then be obtained from

$$\gamma - 1 = \frac{1}{2}v^2/c^2 + \frac{3}{8}v^4/c^4 + 0(v/c)^6 , \tag{5.19}$$

which would yield

$$\begin{aligned} T &= \frac{1}{2}mv^2 + \frac{3}{8}mv^2(v/c)^2 + 0(v/c)^4 \\ &= \frac{1}{2}mv^2\left(1 + \frac{3}{4}v^2/c^2 + 0(v/c)^4\right) . \end{aligned} \tag{5.20}$$

If, instead of $m$ we know $m'$, we have

$$\begin{aligned} T &= (1 - \gamma^{-1})\, m'c^2 \\ &= \left(1 - \sqrt{1 - v^2/c^2}\right) m'c^2 \\ &= \left(\frac{1}{2}v^2/c^2 + \frac{1}{8}v^4/c^4 + 0(v/c)^6\right) m'c^2 \\ &= \frac{1}{2}m'v^2\left(1 + \frac{1}{4}v^2/c^2 + 0(v/c)^4\right) . \end{aligned} \tag{5.21}$$

Another important result is obtained by considering the magnitude of the momentum 4-vector

$$p^\mu p^\nu g_{\mu\nu} = m^2c^2 = (p^0)^2 - \mathbf{p}\cdot\mathbf{p} = E^2/c^2 - p^2 . \tag{5.22}$$

Thus, we can express the energy in terms of the mass and momentum by

$$E^2 = m^2c^4 + p^2c^2 . \tag{5.23}$$

Taking square roots of both sides and bearing in mind that the energy is always positive, the relativistic free-particle Hamiltonian is

$$H = \sqrt{m^2c^2 + p^2}\; c . \tag{5.24}$$

In the case that there is some potential $\phi(\mathbf{x})$, we get

$$H[\mathbf{x},\mathbf{p}] = \sqrt{m^2c^2 + p^2}\; c + \phi(\mathbf{x}) . \tag{5.25}$$

## 2. The Doppler Shift in Relativity

We are familiar with the Doppler effect in acoustics, whereby the sound emitted by an approaching object appears shriller and that of a receding object appears deeper than the sound emitted. In other words there is a velocity dependent shift of the sound frequency. Regarding light as waves, we could also expect a velocity dependent shift of the frequency of light, i.e. a change of colour due to relative motion. In sound the ratio of the observed to the emitted frequency was the sum (difference for the opposite motion) of the velocities of sound and the object divided by the speed of sound. In the case of light we must, obviously, deal with the relativistic addition of velocities and incorporate the effect of time dilation. The resultant calculation would be extremely complicated. What is generally done, to derive the result, is to use another discovery of Einstein (and Planck) which relates the energy of the light to its frequency. It is anyhow obvious that a wave of higher frequency has more energy. What is not so clear is how to relate the energy to the frequency quantitatively.

It was already known, in 1905, that the radiation spectrum could be understood best by regarding electromagnetic radiation as being absorbed or emitted by matter in discrete quanta. Einstein suggested that this could be taken as being due to the electromagnetic radiation consisting of discrete quanta. On this basis he predicted that in the photo-electric effect (emission of electrons from a metal plate due to incident electromagnetic radiation) the current would change with the intensity of the radiation and the voltage with the frequency. This hypothesis was tested and found to be true. The basis for this prediction was the assumption that the energy of a 'wave-packet' of light, a ***photon***, is proportional to the frequency of the light

$$E = h\nu \ , \tag{5.26}$$

where $h$ is known as Planck's constant.

Consider an observer $O$, seeing light emitted by a source $S$ moving with a velocity $\mathbf{v}$ with frequency $\nu'$. Let the motion be along the $x$-direction and, at the instant that the light is emitted, let the

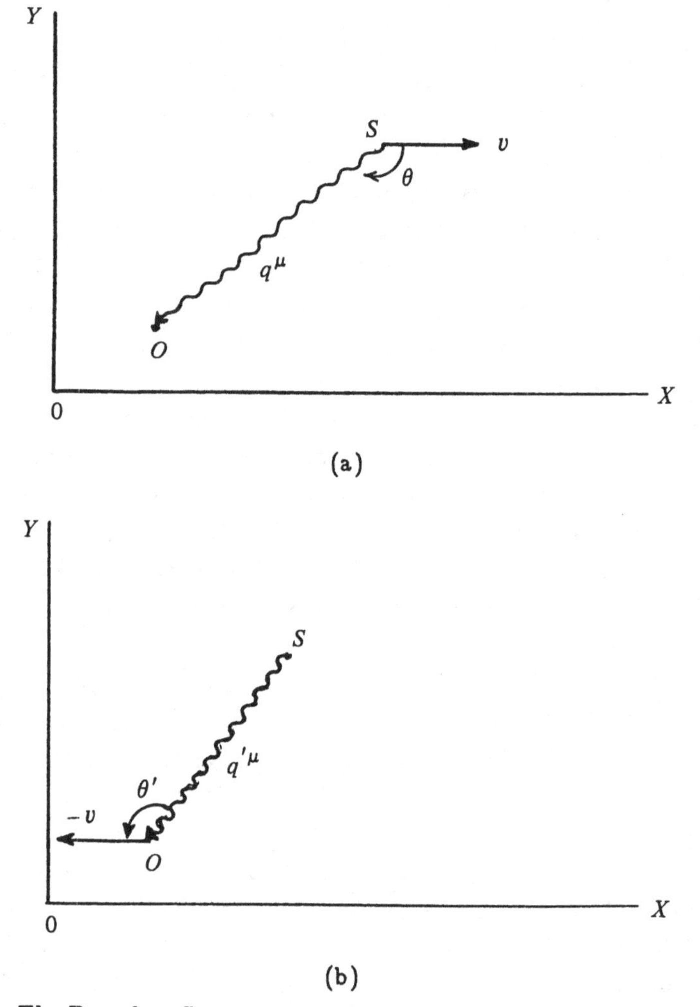


Fig. 18. The Doppler effect causes a change in the observed frequency due to motion of the source $S$, here taken to be in the $x$-direction. The line of sight $OS$ makes an angle $\theta$ with the $x$-axis. (a) is in the frame of $O$ and (b) in the frame of $S$.

line of sight make an angle $\theta$ with the $X$-axis and lie in the $X$ plane (see Fig. 18). Now, from Eqs. (5.14) and (5.26) the rest-mass of the photon is

$$m = \gamma^{-1} h\nu'/c^2 \ . \tag{5.27}$$

Since $\gamma^{-1}$ is zero for light ($\gamma^{-1} = \sqrt{1 - v^2/c^2}$), the photon rest-mass is zero. From Eq. (5.23), then, the energy is just $c$ times the magnitude of the momentum. Let the momentum 4-vector in the $O$-frame be $q^\mu$, given by

$$q^\mu = (q,\ q \cos\theta,\ q \sin\theta,\ 0) \ . \tag{5.28}$$

Applying the Lorentz transformation for motion in the $x$-direction only, given in Eq. (4.12), we see that

$$q'^\mu = \left(\gamma[q - q\frac{v}{c}\cos\theta],\ \gamma[q\cos\theta - q\frac{v}{c}],\ q\sin\theta,\ 0\right) \ . \tag{5.29}$$

Now, from Eq. (5.26), we have

$$q = E/c = h\nu/c \ , \tag{5.30}$$

and using this expression for $q$ in Eq. (5.29), we get

$$\begin{aligned} q' &= E'/c = h\nu'/c \\ &= \gamma q\left(1 - \frac{v}{c}\cos\theta\right) = \gamma\left(1 - \frac{v}{c}\cos\theta\right)\frac{h\nu}{c} \ . \end{aligned} \tag{5.31}$$

Thus, we have the ratio of the emitted to the observed frequency

$$\frac{\nu'}{\nu} = \frac{1 - (v/c)\cos\theta}{\sqrt{1 - v^2/c^2}} \ . \tag{5.32}$$

It is of interest to consider some special cases. For example, if $\theta = \pi$ then

$$\frac{\nu'}{\nu} = \frac{1 + v/c}{\sqrt{1 - v^2/c^2}} = \sqrt{\frac{c+v}{c-v}} > 1 \ , \tag{5.33}$$

i.e. if the motion is radially away from the observer the frequency is decreased. Now, we know that the wavelength is given by

$$\lambda = c/\nu \ . \tag{5.34}$$

Thus, the wavelength is increased. Since blue light has a shorter wavelength and red light a longer wavelength, the shift is towards the red end of the spectrum. It is called a red-shift. If $\theta = 0$, i.e., the motion is towards the observer

$$\frac{\nu'}{\nu} = \frac{1 - v/c}{\sqrt{1 - v^2/c^2}} = \sqrt{\frac{c - v}{c + v}} < 1 \ , \tag{5.35}$$

which gives a blue-shift. If $\theta = \pi/2$, i.e. the motion is perpendicular to the line of sight there is a red-shift

$$\frac{\nu'}{\nu} = 1/\sqrt{1 - v^2/c^2} > 1 \ , \tag{5.36}$$

which is due to time dilation. This would not have been expected on classical considerations but had been observed already. One may ask for what value of $\theta$ will there be no Doppler shift for a given speed of the observer $v$. This is easily obtained by putting $\nu' = \nu$ in Eq. (5.32). Squaring and transposing gives

$$(v^2/c^2) \cos^2 \theta - 2(v/c) \cos \theta + v^2/c^2 = 0 \ , \tag{5.37}$$

which is easily solved to yield

$$\cos \theta = c/v - \sqrt{c^2/v^2 - 1} \ . \tag{5.38}$$

For sufficiently low velocities we get

$$\theta \approx \cos^{-1}(v/2c) \ . \tag{5.39}$$

The Doppler effect is of great importance in the extraction of information about stars from observation of the light coming from them. It is known that every element and compound possesses its own unique spectrum. Due to the Doppler shift the entire spectrum gets shifted. We can work out the speed of a star if we know its Doppler shift and the angle its motion makes with the line of sight. Similarly, we can work out the speed of rotation of galaxies (or stars within galaxies) etc. by the Doppler shift.

## 3. The Compton Effect

Consider light incident on electrons at rest. The light, after encountering the electrons, not only gets dispersed but it also changes colour. This effect is seen, for example, in the reddening of the setting Sun. The wavelength changes. This effect may be understood in relativistic terms by regarding (once again) light as consisting of photons. We can imagine a collision between a single photon and an electron. To start with the electron has the momentum 4-vector $p_1^\mu = (m_e c, \mathbf{0})$ and the photon has the momentum 4-vector $q_1^\mu = (q, q, 0, 0)$. After the collision the electron has the 4-momentum $p_2^\mu = (E/c,\ p\cos\phi,\ -p\sin\phi,\ 0)$ and the photon has the 4-momentum $q_2^\mu = (q',\ q'\cos\theta,\ q'\sin\theta,\ 0)$. This is depicted in

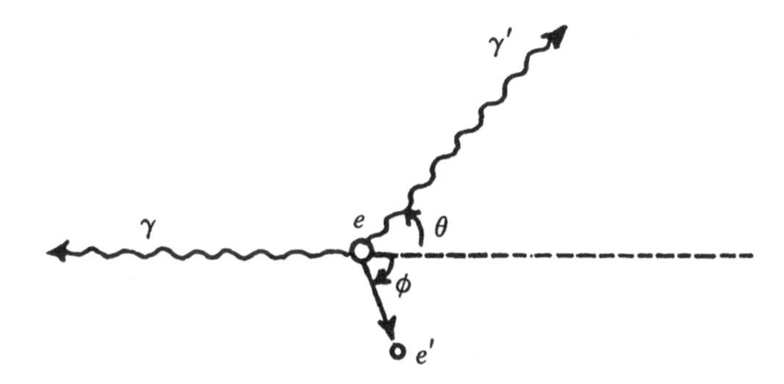


Fig. 19. Compton scattering of light off electrons. At the microscopic level a photon, $\gamma$, of frequency $\nu$ scatters off an electron imparting some momentum to it. The photon goes off at an angle $\theta$ relative to the direction of motion while the electron goes off at an angle $\phi$, measured in the opposite sense to $\theta$.

Fig. 19. Now we notice that, by the conservation of energy and momentum

$$p_1^\mu + q_1^\mu = p_2^\mu + q_2^\mu \ . \tag{5.40}$$

Thus we have

$$\begin{aligned} p_2^\mu p_2^\nu g_{\mu\nu} &= (p_1^\mu + q_1^\mu - q_2^\mu)(p_1^\nu + q_1^\nu - q_2^\nu) g_{\mu\nu} \\ &= p_1^\mu p_1^\nu g_{\mu\nu} + 2p_1^\mu (q_1^\nu - q_2^\nu) g_{\mu\nu} + (q_1^\mu - q_2^\mu)(q_1^\nu - q_2^\nu) g_{\mu\nu} \ . \end{aligned} \tag{5.41}$$

Now, we use the fact that the magnitude squared of the momentum 4-vector is the square of the rest mass multiplied by $c^2$. Thus, the first term on the right in Eq. (5.41) cancels the term on the left. Further, the square of the magnitude of the photon 4-vector momentum is zero. Thus, Eq. (5.41) reduces to

$$2p_1^\mu(q_1^\nu - q_2^\nu)g_{\mu\nu} = 2q_1^\mu q_2^\nu g_{\mu\nu} \ . \tag{5.42}$$

Using the values of $p_1^\mu$, $q_1^\mu$ and $q_2^\mu$ we get

$$m_e c(q - q') = qq' - qq' \cos\theta \ . \tag{5.43}$$

Dividing through by $m_e c\, qq'/h$, we get

$$h/q' - h/q = (h/m_e c)\,(1 - \cos\theta) \ . \tag{5.44}$$

Now, we have

$$h/q' = hc/h\nu' = \lambda',\ h/q = \lambda \ . \tag{5.45}$$

Thus we finally obtain

$$\lambda' = \lambda + \lambda_c\,(1 - \cos\theta) \ , \tag{5.46}$$

where $\lambda_c$ is the Compton wavelength

$$\lambda_c = h/m_e c \ . \tag{5.47}$$

If $\theta = 0$, i.e., when light goes straight, $\lambda' = \lambda$. If the light is deflected at right angles the wavelength is increased exactly by the Compton wavelength. If the light is sent back, $\theta = \pi$, the wavelength is increased by twice the Compton wavelength, $\lambda' = \lambda + 2\lambda_c$. Clearly, $p$, $\phi$ and $E$ can also be determined for any given $\theta$, i.e., the recoiling electron's energy and momentum are determinable.

The result is directly obtainable from Eq. (5.40) for $\mu = 1, 2$:

$$p\cos\phi = q - q'\cos\theta \ , \tag{5.48}$$

$$p\sin\phi = q'\sin\theta \ . \tag{5.49}$$

Dividing Eq. (5.49) by Eq. (5.48), we get

$$\tan\phi = q'\sin\theta/(q - q'\cos\theta)\ . \tag{5.50}$$

Dividing through by $qq'$ and using Eq. (5.45), we get

$$\tan\phi = \lambda\sin\theta/(\lambda' - \lambda\cos\theta)\ . \tag{5.51}$$

Now, using Eq. (5.46), we get

$$\phi = \tan^{-1}\left(\frac{\sin\theta}{(1+\lambda_c/\lambda)(1-\cos\theta)}\right) = \tan^{-1}\left(\frac{\cot(\theta/2)}{(1+\lambda_c/\lambda)}\right)\ . \tag{5.52}$$

In the case that $\lambda_c \ll \lambda$ we get the classical result

$$\phi \approx \tan^{-1}\left(\sin\theta/(1-\cos\theta)\right) = \tan^{-1}(\cot\theta/2)\ , \tag{5.53}$$

while, if $\lambda_c \gg \lambda$ we get

$$\phi \approx \tan^{-1}\left(\lambda\sin\theta/\lambda_c(1-\cos\theta)\right) = \tan^{-1}\left((\lambda/\lambda_c)\cot(\theta/2)\right)\ . \tag{5.54}$$

If $\theta = 0$, clearly $\phi = \pi/2$ and if $\theta = \pi$, $\phi$ is zero. If $\theta = \pi/2$

$$\phi = \tan^{-1}\left(\lambda/(\lambda+\lambda_c)\right)\ . \tag{5.55}$$

Now, squaring Eqs. (5.48), and (5.49) adding and taking square roots gives

$$\begin{aligned} p &= \sqrt{(q-q'\cos\theta)^2 + q'^2\sin^2\theta} \\ &= q'\sqrt{1-(2q/q')\cos\theta + (q/q')^2} \\ &= \frac{h}{\lambda+\lambda_c(1-\cos\theta)} \\ &\quad\times\sqrt{1-2[1+(1-\cos\theta)\lambda_c/\lambda]\cos\theta + [1+(1-\cos\theta)\lambda_c/\lambda]^2}\ . \end{aligned} \tag{5.56}$$

Further, the energy of the electron is given by

$$\begin{aligned} E &= \sqrt{p^2c^2 + m_e^2c^4} \\ &= pc\sqrt{1+\lambda_e^2 m_e^2 c^2/h^2} \\ &= pc\sqrt{1+\lambda_e^2/\lambda_c^2}\ , \end{aligned} \tag{5.57}$$

where $p$ is given by Eq. (5.56) and $\lambda_e$ is the De Broglie wavelength of the electron. Thus, we have a complete description of the scattering of the photon through an angle $\theta$ by an electron at rest relative to the observer, initially.

## 4. Particle Scattering

As mentioned earlier, kinematics though invented earlier came into its own with the advent of Relativity. Most importantly it was developed for the relativistic theory of scattering. Generally we deal with two particles which are initially independent of each other coming closer together, interacting and moving away till they are again independent of each other. There could be long-range forces acting between them. In mechanics, as dealt with here, we only deal with contact forces and *no* long-range forces. In this case the two particles must collide. We deal with perfectly elastic collisions so that momentum may be transferred from one particle to another but cannot be dissipated. In Relativity we can transform from one frame to another *and derive physically significant results*, unlike the classical case.

We have already seen the result of scattering of a massless particle (a photon) off a massive particle (an electron). In that case it was convenient to use the frame in which the massive particle was initially at rest. This frame was more convenient because it was the frame in which the observer performing the experiment would be. Generally we call this frame the 'laboratory frame'. It is necessary to state the final result in terms of the laboratory frame, where the results would be tested. However, it is often convenient to perform the calculations in the so-called centre of mass frame. Given the momentum 4-vectors of both particles, the centre of mass frame is the one in which the total momentum 4-vector, which is the sum of the two 4-vector momenta of the particles, has zero spatial components. We need to be able to find the Lorentz transformation relating the laboratory frame to the centre of mass frame.

Consider two particles having 4-vector momenta in the laboratory frame, $p_1^\mu$ and $p_2^\mu$. Thus, the total momentum 4-vector is

$$P^\mu = p_1^\mu + p_2^\mu = (E/c, \mathbf{P}) \ . \tag{5.58}$$

In the centre of mass frame the momenta would be $p_1^{'\mu}$, $p_2^{'\mu}$ and

$$P^{'\mu} = p_1^{'\mu} + p_2^{'\mu} = (E'/c, \mathbf{0}) \ . \tag{5.59}$$

Now, from Eqs. (5.7), (5.8) and (5.14) it is obvious that

$$\mathbf{P}/E = \mathbf{V}/c^2 \ . \tag{5.60}$$

In the centre of mass frame, of course, the velocity is zero. Thus $\mathbf{V}$ is the relative velocity between the centre of mass frame and the laboratory frame. Clearly we have

$$p_1' = -p_2' = p' \quad \text{(say)} \ . \tag{5.61}$$

If the $x$-direction is taken to be along $\mathbf{V}$, we have

$$\left.\begin{aligned} p_{1x} &= \frac{p_x' + E_1'V/c^2}{\sqrt{1-V^2/c^2}} \ , \quad p_{1y} = p_y', \ p_{1z} = p_z' \ ; \\ p_{2x} &= \frac{-p_x' + E_2'V/c^2}{\sqrt{1-V^2/c^2}} \ , \ p_{2y} = -p_y', \ p_{2z} = -p_z' \ ; \\ E_1 &= \frac{E_1' + p_x'V}{\sqrt{1-V^2/c^2}} \ , \quad E_2 = \frac{E_2' - p_x'V}{\sqrt{1-V^2/c^2}} \ . \end{aligned}\right\} \tag{5.62}$$

For example, consider a particle of rest-mass $m_1$ and velocity, in the laboratory frame $\mathbf{v}$, colliding with a particle of rest-mass $m_2$ at rest in the laboratory frame. Now

$$\mathbf{p}_1 = \frac{m_1\mathbf{v}_1}{\sqrt{1-\mathbf{v}_1\cdot\mathbf{v}_1/c^2}} \ , \quad \mathbf{p}_2 = \mathbf{0} \ . \tag{5.63}$$

Thus the total momentum 4-vector is

$$P^\mu = (m_2c + m_1c/\sqrt{1-\mathbf{v}_1\cdot\mathbf{v}_1/c^2}, \ m_1\mathbf{v}_1/\sqrt{1-\mathbf{v}_1\cdot\mathbf{v}_1/c^2}) \ . \tag{5.64}$$

Hence we have

$$\mathbf{V} = \left( \frac{m_1/\sqrt{1 - \mathbf{v}_1 \cdot \mathbf{v}_1/c^2}}{m_1/\sqrt{1 - \mathbf{v}_1 \cdot \mathbf{v}_1} + m_2} \right) \mathbf{v}_1 \ . \tag{5.65}$$

Or, more simply

$$\mathbf{V} = \frac{m_1 \mathbf{v}_1}{m_1 + m_2 \sqrt{1 - \mathbf{v}_1 \cdot \mathbf{v}_1/c^2}} \ , \tag{5.66}$$

gives the velocity of the centre of mass frame relative to the laboratory frame. In this frame the Lorentz transformation of $p_2^\mu$ is given by

$$p_2^{'\mu} = \left( \frac{m_2 c}{\sqrt{1 - \mathbf{V} \cdot \mathbf{V}/c^2}} \ , \ \frac{-m_2 \mathbf{V}}{\sqrt{1 - \mathbf{V} \cdot \mathbf{V}/c^2}} \right) \ . \tag{5.67}$$

Now, since

$$\mathbf{p}_1' + \mathbf{p}_2' = 0 \ , \tag{5.68}$$

we have

$$\mathbf{p}_1' = \frac{m_2 \mathbf{V}}{\sqrt{1 - \mathbf{V} \cdot \mathbf{V}/c^2}} \ . \tag{5.69}$$

Also, the Lorentz transformation of the energy gives

$$\begin{aligned} E_1' &= \frac{E_1 - \mathbf{p}_1 \cdot \mathbf{V}}{\sqrt{1 - \mathbf{V} \cdot \mathbf{V}/c^2}} \\ &= \frac{m_1 c^2 (1 - \mathbf{v}_1 \cdot \mathbf{V}/c^2)}{\sqrt{1 - \mathbf{v}_1 \cdot \mathbf{v}_1/c^2} \sqrt{1 - \mathbf{V} \cdot \mathbf{V}/c^2}} \end{aligned} \tag{5.70}$$

which finally yields

$$p_1^{'\mu} = \left( \frac{m_1 c (1 - \mathbf{v}_1 \cdot \mathbf{V}/c^2)}{\sqrt{1 - \mathbf{v}_1 \cdot \mathbf{v}_1/c^2} \sqrt{1 - \mathbf{V} \cdot \mathbf{V}/c^2}} \ , \ \frac{m_2 \mathbf{V}}{\sqrt{1 - \mathbf{V} \cdot \mathbf{V}/c^2}} \right) \ . \tag{5.71}$$

Equations (5.66), (5.67) and (5.71), between them, give the four-momenta in the centre of mass frame.

Notice that relativistically, even if $m_1 = m_2$, i.e. the particles are of equal mass, the centre of mass frame velocity is not simply half the velocity of the projectile. However, in the limit $c \to \infty$ we recover the classical kinematic results as Eq. (5.66) now becomes

$$\mathbf{V} = m_1 \mathbf{v}_1/(m_1 + m_2) \ , \tag{5.72}$$

which gives $\mathbf{V} = \mathbf{v}_1/2$ when $m_2 = m_1$. Also, the momenta given by Eqs. (5.67) and (5.71) now simply become

$$\mathbf{p}_1' = m_2 \mathbf{V} \, , \ \mathbf{p}_2' = -m_2 \mathbf{V} \ . \tag{5.73}$$

Using Eq. (5.72) we see that

$$\mathbf{p}_1' = m_1 m_2 \mathbf{v}_1/(m_1 + m_2) = -\mathbf{p}_2' \ , \tag{5.74}$$

yielding the usual formula for the reduced mass

$$\mu = m_1 m_2/(m_1 + m_2) \ . \tag{5.75}$$

## 5. Binding Energy, Particle Production and Particle Decay

One of the most important consequences of relativistic kinematics is the equivalence of mass and energy. At about the time that the formula given by Eq. (5.14) was presented, the phenomenon of radio-activity in radium had been observed. The question arose as to the source of the energy emitted by radio-activity. Then, too, with the advent of the statistical interpretation of thermodynamics it became important to understand the physical nature of the chemical potential, which is a very important thermodynamic quantity. Further, there was no way to understand why the Sun shines. The sheer magnitude of the solar energy source made all known energy generation mechanisms inadequate.

The most favoured theory was that the solar material was not in equilibrium, but was collapsing inwards. The observed amount of solar energy was, then, the kinetic energy of the falling solar matter being converted into radiation by the heating up of the solar gas. To provide the observed energy output of the Sun, the solar matter would have to fall in at a rate of about 30 km per year. This decrease in size was undetectable by the best experiments available at the time. However, for the same amount of energy to have been produced steadily over very long periods it would be necessary to suppose that the Sun's size was equal to the Earth's orbit some millions of years ago. According to this theory, then, either the Sun started shining so strongly only recently or the Earth should be only some millions of years old. Now geological and palaeontological evidence proved that the Sun had been shining on the Earth as brightly as it does today for billions of years! Thus the theory had to be wrong.

It is now understood that chemical energy comes from breaking (or otherwise modifying) the electronic structure of atoms. The Sun shines due to changes in the structure of nuclei and radio-activity is essentially due to sub-nuclear forces causing changes of the nuclear (or sub-nuclear) structure. In the first case the energy source can be the electrostatic potential between the charged nucleus and the atomic electrons. However, no such energy storing mechanism could be conceived for the nuclear energy release. The equivalence of mass and energy provided a possible answer to this problem. ***Energy was stored as mass.*** For example the solar energy produced over billions of years would produce a negligible change in the Sun's mass. However, a mechanism was required to convert mass into energy. For the Sun a partial explanation was provided by Hans Bethe in 1956. It should be mentioned that as of even date the theory still has problems since it makes some predictions which are not found to hold true. However, it is clear that whatever the details, mass is being converted into energy all the time. This fact has been tested. For heavy nuclei, the sum of the masses of two fission fragments is less than the mass of the original nucleus. For light nuclei, on the other hand, the sum of the masses of the fusing nuclei is greater than the mass of the resultant nucleus. The difference of masses is called the ***mass defect***. This mass defect gives a measure of the difference of ***binding energies*** for the nuclei, i.e. the potential energy of the nucleus being bound. Essentially, the difference of binding energies between nuclei is equal to the corresponding (fission or fusion) mass defect times the square of the speed of light. The binding energy of nuclei, as determined presently, is given in Fig. 20.

Of particular importance, among the consequences of the equivalence of mass and energy, is the expectation that energy could be converted into mass, not in the sense of increasing the moving mass of a single particle, but by creating more rest-mass in the form of a new particle. The picture is that one particle would strike another and move off. The target particle would also move off. In addition,

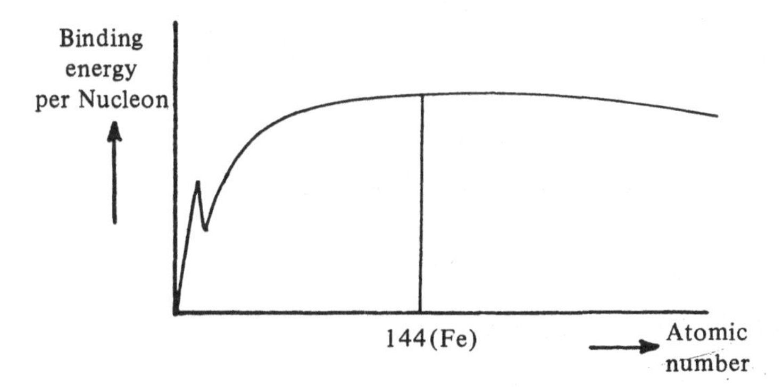


Fig. 20. The binding energy of nuclei plotted against the atomic number gives a peak at the atomic number for iron. Fusing lower elements together (nuclear fusion) to form elements up to iron or breaking larger elements down (nuclear fission) releases the binding energy.

a new particle would be produced. Let us denote the rest-mass of the three particles by $m_P$ (for the projectile), $m_T$ (for the target) and $m_N$ (for the new particle produced). Given these masses we could ask what the minimum kinetic energy for particle production would be. In the laboratory frame let the projectile have an energy $E$. Then the 4-vector momenta of the projectile and the target particles would be

$$p_P^\mu = (E/c, \mathbf{p}), \;\; p_T^\mu = (m_T c, \mathbf{0}) \; . \tag{5.76}$$

Thus, the total 4-vector momentum before collision is

$$P^\mu = (E/c + m_T c, \, \mathbf{p}) \; . \tag{5.77}$$

Thus, we have

$$P^{\mu} P^{\nu} g_{\mu\nu} = (E/c + m_T c)^2 - |\mathbf{p}|^2 \ . \tag{5.78}$$

From Eq. (5.23) we know that

$$E^2/c^2 - |\mathbf{p}|^2 = m_P^2 c^2 \ , \tag{5.79}$$

which gives the invariant square of the total energy

$$P^{\mu} P^{\nu} g_{\mu\nu} = (m_P^2 + m_T^2)c^2 + 2m_T E \ . \tag{5.80}$$

We could also calculate the transformation to the centre of mass frame, as discussed earlier. The total momentum 4-vector in that frame is

$$P'^{\mu} = (W/c, \mathbf{0}) \ , \tag{5.81}$$

where $W$ is the total energy of the system. Now, due to Lorentz invariance

$$P'^{\mu} P'^{\nu} g_{\mu\nu} = P^{\mu} P^{\nu} g_{\mu\nu} \ . \tag{5.82}$$

Using Eqs. (5.80) and (5.81) in Eq. (5.82) gives the equation

$$W^2/c^2 = (m_P^2 + m_T^2)c^2 + 2m_T E \ . \tag{5.83}$$

For particle production to occur, the total energy must exceed the final rest-energy

$$m_f c^2 = (m_P + m_T + m_N)c^2 \ . \tag{5.84}$$

The kinetic energy of the projectile, $T$, is then given by

$$\begin{aligned} T &= E - m_P c^2 \\ &= (W^2 - m_i^2 c^4)/2m_T c^2 \ , \end{aligned} \tag{5.85}$$

where $m_i$ is the initial mass

$$m_i = (m_P + m_T) \ . \tag{5.86}$$

The threshold energy, $T_0$, which is the least kinetic energy that can lead to particle production, is given by putting

$$W^2 = m_f^2 c^4 \ , \tag{5.87}$$

in Eq. (5.85). Thus we get

$$T_0 = (m_f^2 - m_i^2)c^2/2m_T \ . \tag{5.88}$$

Notice that, as far as these calculations are concerned, the number of new particles produced does not matter – $m_N$ could equally well stand for the sum of the rest-masses of any number of particles. If $T > T_0$ the energy of motion will get disturbed among the resultant particles. Therefore, there would not be any unique way that the resultant particles could move. Instead, there would be a range of possible outcomes. The particular outcome could be worked out only on the basis of some additional information.

Relativity also provides for the reverse of particle production – particle decay. In this case a particle of total mass $M$, travelling along with a momentum 4-vector

$$P^\mu = (E/c, \mathbf{P}) \ , \tag{5.89}$$

suddenly breaks into two particles of rest-masses $m_1$ and $m_2$ and momentum 4-vectors

$$p_1^\mu = (E_1/c, \mathbf{p}_1), \ \ p_2^\mu = (E_2/c, \mathbf{p}_2) \ . \tag{5.90}$$

In the rest-frame of the initial particle the momentum is zero. Thus, the centre of mass frame of the final particles is the rest-frame of the initial particle. It has a velocity relative to the laboratory frame given by

$$\mathbf{v} = \mathbf{P}c^2/E \ . \tag{5.91}$$

In that frame the initial momentum 4-vector is

$$P'^\mu = (Mc, \mathbf{0}) \ . \tag{5.92}$$

This must be the final momentum 4-vector, due to momentum conservation. The individual 4-vector momenta are

$$p_1'^{\mu} = (E_1'/c, \mathbf{p}_1'),\ p_2'^{\mu} = (E_2'/c, \mathbf{p}_2') \ . \tag{5.93}$$

Since we also have

$$p_1'^{\mu} + p_2'^{\mu} = P'^{\mu} \ , \tag{5.94}$$

we have

$$\mathbf{p}_1' = -\mathbf{p}_2' = \mathbf{p} \quad \text{(say)} \ , \tag{5.95}$$

$$E_1' + E_2' = Mc^2 \ . \tag{5.96}$$

If the rest-masses of the decay products are $m_1$ and $m_2$

$$E_1' = \sqrt{p^2 + m_1^2c^2}\ c, \quad E_2' = \sqrt{p^2 + m_2^2c^2}\ c \ . \tag{5.97}$$

Since there are three equations for the six parameters $M$, $p$, $E_1'$, $E_2'$, $m_1'$ and $m_2'$, we need to know three of them to determine the other three. Once these are worked out, the relevant 4-vectors can be transformed by the Lorentz transformation given by $\mathbf{v}$ as specified by Eq. (5.91).

We could have a decay of one particle into more than two constituents. This happens, for example in the decay of neutrons,

$$n \to p + e + \overline{\nu}_e \ , \tag{5.98}$$

where $p$ represents the proton, $e$ the electron and $\overline{\nu}_e$ is the antineutrino. In fact the original indication that there were particles such as the neutrino came from the application of the above calculations to what was supposed to be the decay process: the neutron decaying into a proton and an electron. Both particles were observed and their energy worked out. There was an energy defect between the rest-energy of the neutron and the total energy of the proton and the electron. A particle was postulated which carried away the extra energy. Since the neutron is neutral and the proton and electron have equal and opposite charges, the postulated particle, the anti-neutrino, must also be neutral. Neutrinos have, since, been detected experimentally and are used regularly in many experiments nowadays.

## Exercise 5

1. A beam of protons is incident on a target of Carbon. Taking the atomic number of the Carbon to be exactly 12 (consisting of six protons and six neutrons) and the masses of protons and neutrons to be 938.2 MeV$/c^2$ and 939.5 MeV$/c^2$, respectively, determine the threshold energy for the production of pions of mass 140 MeV$/c^2$.

2. Two bodies have the same rest-mass and are moving at a speed $c/2$ relative to each other when they collide. What is the amount of energy transferred according to an observer at rest relative to one of the two bodies? If the observer sees them moving co-linearly and sees them both coming towards him, one of them with speed $c/3$, what will the speed of the other body seem to him and how much energy transfer will he see?

3. $A$ sees a body of rest-mass $m$ as having energy $2mc^2$ while $B$ sees its energy as $mc^2$. $B$ sees a body of rest-length $\ell$, in the frame of $C$, as having a length $\ell/3$. $C$ sees a clock in the rest-frame of $D$ as running 4-times too slow. Taking the object in the frame of $C$ as lying along the direction of relative motion between $B$ and $C$, give the possible velocities of $D$ relative to $A$: (i) if all the observers have relative motion in the same direction; (ii) if the motions are all perpendicular; and (iii) if one motion is perpendicular to the other two which are co-linear. (Remember that there can be two signs in each case.)

4. If an object moves, relative to an observer, at a speed $c/2$, what should be the angle between the velocity vector and the line

joining the object and the observer, so that no Doppler shift is observed?

5. An object is in a circular orbit about an observer. Neglecting acceleration, what is the Doppler shift seen by the observer if the radius of the orbit is $R$ and the mass of the gravitating source (on which the observer stands) is $M$?

6. What should the wavelength of light be so that the scattered light has double the wavelength of the original light when scattered through angles of: (i) $\varepsilon \approx 0$; (ii) $\pi/6$; (iii) $\pi/4$; (iv) $\pi/3$; (v) $\pi/2$; (vi) $2\pi/3$; (vii) $3\pi/4$; (viii) $5\pi/6$; and (ix) $\pi$? Take $m_e = 9.11 \times 10^{-31}$ kg, $h = 6.63 \times 10^{-34}$ J.sec, $c = 3 \times 10^8$ m/sec.

7. A proton (of rest-mass 938.2 MeV/$c^2$) strikes another proton and produces two new particles of rest-masses 1 346 MeV/$c^2$ and 3 154 MeV/$c^2$ in addition to the original protons. What is the least energy that the incoming proton must have had?

8. The Glashow-Salam-Weinberg theory predicted the existence of a $Z^\circ$ of rest-mass 90 000 MeV/$c^2$. These particles were to be produced by proton-proton collisions by accelerating the protons, splitting the beam and bringing the protons to a head-on collision. How much energy did the accelerator have to accelerate the protons to? How can this energy requirement be reduced?

9. In the particle reaction

$$\mu \rightarrow e + \overline{\nu}_e + \nu_\mu \ ,$$

if the neutrinos, $\overline{\nu}_e$ and $\nu_\mu$ are nearly massless and $m_e = 0.51$ MeV/$c^2$ while $m_\mu = 105$ MeV/$c^2$, what limits can be put on the speed, momentum and energy of the electrons? If the neutrinos are not massless, but have a small rest-mass of about 50 eV/$c^2$ (1 eV $= 10^{-6}$ MeV) each, how accurately would we have to measure the energy of the electron for the neutrino mass to be detectable?

10. An observer, $A$, sees a body of rest-mass $m$ as having mass $M$. An observer, $B$, comoving with that body sees another body of rest-length $\ell$ as having length $L$. An observer $C$, comoving with this other body sees $A$'s watch as going at half the correct rate. If the motion is co-linear and $A$'s watch is working correctly in his frame, determine $L$ in terms of $\ell$, $m$ and $M$.

11. Two meteors of rest-mass 0.1 and 0.2 kg, respectively, collide. If the relative speed before collision is 0.1 $c$ and an observer sees them coming with equal and opposite speed and sees the lighter meteor go off at right angles to the original direction of motion, what will be the deflection of the heavier meteor according to the observer? How will this process appear to an observer comoving with: (i) the heavier meteor; and (ii) the lighter meteor? How will it appear to the centre of mass observer?

12. X sees a spaceship coming towards him at a speed such that the length of the spaceship seems to have been halved. He sees a meteor of 1 g coming towards him from the opposite direction at half the speed of light. He wants to warn the spaceship about the effective mass and momentum of the meteor as it will appear to the spaceship. What values should he give?

13. A spaceship of rest-mass 10 000 kg is hit by a micro-meteor of rest-mass 0.1 kg travelling at a speed of 0.99 $c$ relative to the Earth at right angles to the motion of the liner. The meteor is deflected off the spaceship's shields. If the spaceship was travelling at a speed of 100 000 km/s relative to the Earth, how much is it deflected from its original course? If no corrections are made, how far off target will the spaceship arrive at its destination $4\frac{1}{2}$ light years from the collision?

14. A particle of rest-mass 3 GeV/$c^2$ (1 GeV = 1 000 MeV) strikes a particle of 1 GeV/$c^2$ mass, at rest in the laboratory, and produces a particle of rest-mass 2 GeV/$c^2$ along with the original particles.

What threshold energy for this process is seen by an observer moving at a speed $2c/5$ relative to the laboratory?

15. An astronomer sees the colour of a star change over a period of 5 hours, roughly corresponding to a change of frequency by a factor of 2. How massive is the companion star of the visible star if they both have equal masses. How much "wobble" must there be due to the gravitational pull of the normal star (which is visible) on its companion, if the normal star has one solar mass ($2 \times 10^{30}$ kg)? You may take $c = 3 \times 10^5$ km/s and $G = 6.67 \times 10^{-11}$ Nm$^2$/kg$^2$. (The "wobble" is the radius of the orbit of the companion star.)

16. The light from a star $10^{13}$ km away is red-shifted by 10% and the star is seen to move across the sky at a rate of $10^{-3}$ radians per year. Determine the velocity of the star.

17. An electron is moved from rest by a photon which is deflected by $\pi/3$ and whose wavelength is doubled. Determine the resultant momentum of the electron.

18. Two particles of equal rest-mass have a head-on collision in the laboratory frame, where one particle has twice the momentum of the other. Determine the centre of mass frame.

19. If the wavelength of a photon increases by 50% on being reflected by an electron, how much will it be altered on scattering off a proton through an angle $\pi/6$? Take the proton to be 1840 times as massive as an electron.

## Chapter 6

# ELECTROMAGNETISM IN SPECIAL RELATIVITY

### 1. Review of Electromagnetism

The theory of Electromagnetism as developed by Maxwell and modified by Lorentz (to incorporate sources of the electromagnetic field) already led Lorentz to present his transformations as *ad hoc* assumptions. We have seen that all of the results of the Special Theory of Relativity follow from the Lorentz transformations. Even without Einstein, his theory would have been forced on us by Electromagnetism. The question arises, then, of whether Electromagnetism is itself Lorentz covariant, i.e. whether the theory is consistent with Special Relativity *without any modification*? To look into this question we shall briefly review the theory as finally formulated by Maxwell, in modern notation. It will be seen that it can be cast into 4-vector form and gives the usual results under Lorentz transformations. In fact, a deeper understanding is obtained by recasting Maxwell's theory in Special Relativistic terms.

The Ancient Greeks knew (and perhaps the Ancient Chinese as well) that there are non-gravitational forces. The force exerted by substances which had been rubbed against other (particular) substances was the *electric* force. Of course, in those days, there was no clear concept of forces in the Newtonian sense. After much experimentation, Coulomb was able to formulate the force due to stationary electric charges acting on each other. This can be written in modern notation as

$$\mathbf{F}_{12} = \frac{1}{4\pi\varepsilon} \frac{q_1 q_2 (\mathbf{r}_1 - \mathbf{r}_2)}{|\mathbf{r}_1 - \mathbf{r}_2|^3} \ . \tag{6.1}$$

Here $\varepsilon$ is a constant depending on the nature of the medium in which the charges $q_1$ and $q_2$ are placed at the points given by position vectors $\mathbf{r}_1$ and $\mathbf{r}_2$ respectively. This law is known as the *Coulomb force law* for *electrostatic forces.* Notice the similarity between this law and Newton's law of gravitation. The constant in that case is $G$ instead of $1/4\pi\varepsilon$ and we have masses $m_1$, $m_2$ instead of charges $q_1$, $q_2$. The difference is that $G$ is a Universal constant but $\varepsilon$ depends on the nature of the medium. Also all bodies have masses which are positive while there can be electrically neutral bodies, i.e., having zero charge, and there can be attraction as well as repulsion.

A single charged particle is called an *electric monopole.* An electric *dipole* is constructed by two oppositely charged monopoles held a fixed distance apart. Whereas no magnetic analogue of the electric monopole has been discovered, a bar magnet is a *magnetic dipole.* The strength of the magnetic poles will be represented by $\mathcal{M}_1$ and $\mathcal{M}_2$. Then *Coulomb's law for magnetostatics* is

$$\mathbf{F}_{12} = \frac{\mu}{4\pi} \frac{\mathcal{M}_1 \mathcal{M}_2 (\mathbf{r}_1 - \mathbf{r}_2)}{|\mathbf{r}_1 - \mathbf{r}_2|^3} \ , \tag{6.2}$$

where $\mu$ is a constant which depends on the magnetic nature of the material. The constants $\varepsilon$ and $\mu$ are called, respectively, the *dielectric constant* and *magnetic permitivity,* of the medium. They both have finite values in vacuum denoted by $\varepsilon_0$ and $\mu_0$ respectively. It should be pointed out that since no magnetic monopoles have been seen, Coulomb's law is to be taken as an approximation for two very long bar magnets with one pole each put relatively close together. Effectively, they would interact as monopoles.

## 2. The Electric and Magnetic Field Intensities

If a bar magnet is placed on a sheet of paper, putting iron filings on the paper enables us to trace the lines of force about the magnet. Of course, the iron filings themselves change the lines of force. The less (and the smaller) the filings, the less do they affect the lines of force of the magnet *by itself.* To get the lines due to the magnet *with no interference* by anything else, the filings would have to be infinitesimal. The force exerted on such infinitesimal objects, called *test particles*, divided by the magnetic (electric charge for the electrostatic case) pole strength of the test particle is called the *magnetic (electric)* field intensity. We can write the electric field intensity due to a point charge, $Q$, as

$$\mathbf{E} = \lim_{q\to 0} \frac{\mathbf{F}_q}{q} = \frac{Q}{4\pi\varepsilon_0}\frac{\mathbf{r}}{r^3} , \tag{6.3}$$

where $\mathbf{r}$ is the position vector of the infinitesimal charge $q$ and $r$ is its magnitude. Similarly, the magnetic field intensity is

$$\mathbf{H} = \lim_{m\to 0} \frac{\mathbf{F}_m}{m} = \frac{\mu_0 \mathcal{M}}{4\pi}\frac{\mathbf{r}}{r^3} . \tag{6.4}$$

In media with electric and magnetic properties, the constants $\varepsilon_0$ and $\mu_0$ can be replaced by $\varepsilon$ and $\mu$ provided that the media are isotropic (i.e., they do not alter the direction of the field vectors).

Now, if there is a collection of charges, or a continuous charge density, it is seen by using Gauss' theorem that

$$\nabla\cdot\mathbf{E} = \rho/\varepsilon_0 , \tag{6.5}$$

where $\rho$ is the charge density. For discrete charges $q_i$, we have

$$\nabla\cdot\mathbf{E} = \Big(\sum_i q_i\,\delta(\mathbf{r}-\mathbf{r}_i)\Big)/\varepsilon_0 , \tag{6.6}$$

where $\delta(\mathbf{r}_i)$ is the Dirac delta function as a function of the position vector, $\mathbf{r}_i$ being the position vector of the $i^{\text{th}}$ particle. The Dirac delta function is defined to be zero if its argument is non-zero and infinite if the argument is zero, such that for a domain $\mathcal{D}$

$$\left.\begin{aligned} \int_{\mathcal{D}} \delta(\mathbf{r}-\mathbf{r}_0)\, f(\mathbf{r})\, dV &= f(\mathbf{r}_0) \quad \text{if } \mathbf{r}_0 \in \mathcal{D} , \\ &= 0 \qquad\quad \text{if } \mathbf{r}_0 \notin \mathcal{D} . \end{aligned}\right\} \tag{6.7}$$

It is easily verified that the volume integral of both Eqs. (6.5) and (6.6) is the total charge over $\varepsilon_0$. By Gauss' theorem the left side of both equations is

$$\int_D \nabla \cdot \mathbf{E}\, dV = \oint \mathbf{E} \cdot \mathbf{n}\, dS \ . \tag{6.8}$$

Using Eq. (6.3) to insert the value of **E**, we see that

$$\int_D \nabla \cdot \mathbf{E}\, dV = (Q/4\pi\varepsilon_0) \oint (\mathbf{r} \cdot \mathbf{n}/r^3)\, dS \ . \tag{6.9}$$

Now the integral is known to be a solid angle $4\pi$. Thus, we see that the volume integral of the left side of either of Eqs. (6.5) and (6.6) is also $Q/\varepsilon_0$. Hence we get *Gauss' law* given by Eq. (6.5) for a continuous charge distribution or Eq. (6.6) for a discrete charge distribution.

In the case of magnetism, since there are no known magnetic monopoles, there is no meaningful charge distribution to put on the right side of Eqs. (6.5) and (6.6). The only way to have a net pole strength inside a volume is to keep half the magnet outside it. Thus, the total strength would not remain unchanged by slight distortions of the volume. Hence, we must keep the right side zero, i.e., *Gauss' law for magnetic fields* is

$$\nabla \cdot \mathbf{H} = 0 \ . \tag{6.10}$$

## 3. The Electric Current

A moving charge gives a *current*. The larger or faster the charge, the greater the current. For a collection of charges $q_\alpha$ moving with velocity $\mathbf{v}_\alpha$ across an area element $d\mathbf{S}$ in time $\delta t$, the current differential will be

$$dI = \delta Q/\delta t = \sum_\alpha q_\alpha \mathbf{v}_\alpha \cdot d\mathbf{S}\, \delta t/\delta t = \sum_\alpha q_\alpha \mathbf{v}_\alpha \cdot d\mathbf{S} \ . \tag{6.11}$$

We write it in terms of the *current density*, $\mathbf{j}$

$$dI = \mathbf{j} \cdot d\mathbf{S} \ . \tag{6.12}$$

The net inflow of current *into* a volume $\mathcal{D}$ bounded by a surface $S$ is, then (by using Gauss' theorem)

$$-I = \oint_S \mathbf{j} \cdot d\mathbf{S} = \int_{\mathcal{D}} \nabla \cdot \mathbf{j} \, dV \ . \tag{6.13}$$

On the other hand the inflow is given directly in terms of the rate of charge flow. For a continuous charge distribution

$$I = dQ/dt = d/dt \left( \int_{\mathcal{D}} \rho \, dV \right) \ . \tag{6.14}$$

The derivative can be taken inside the integral, bearing in mind that there is only differentiation of the charge density with respect to time and not position. Thus, we have a partial and not a total derivative with respect to time. Adding Eqs. (6.13) and (6.14), we get

$$\int_{\mathcal{D}} (\partial \rho / \partial t + \nabla \cdot \mathbf{j}) \, dV = 0 \ . \tag{6.15}$$

Now the integral is generally zero only if the integrand is zero, which yields

$$\partial \rho / \partial t + \nabla \cdot \mathbf{j} = 0 \ . \tag{6.16}$$

This is called the *equation of continuity*.

Two observational laws relating the electric and magnetic field intensities were stated by Ampere and Faraday. They could be written in modern notation as

$$\nabla \wedge \mathbf{H} = \mathbf{j} + \varepsilon_0 \, \partial \mathbf{E} / \partial t \ , \tag{6.17}$$

$$\nabla \wedge \mathbf{E} = -\mu_0 \, \partial \mathbf{H} / \partial t \ . \tag{6.18}$$

Also, consistent with Eqs. (6.5) and (6.10), we can define the *electric scalar potential*, $\phi$, such that

$$\mathbf{E} = -\nabla \phi \tag{6.19}$$

and the *magnetic vector potential*, $\mathbf{A}$, such that

$$\mathbf{H} = (1/\mu_0) \nabla \wedge \mathbf{A} \ . \tag{6.20}$$

Equation (6.20) obviously implies Eq. (6.10) as an identity since

$$\nabla \cdot (\nabla \wedge \mathbf{A}) = (\nabla \wedge \nabla) \cdot \mathbf{A} = 0 \ . \tag{6.21}$$

Equation (6.3) implies that for a point charge

$$\phi = (1/4\pi\varepsilon_0)\, Q/r \ . \tag{6.22}$$

## 4. Maxwell's Equations and Electromagnetic Waves

Rather than proceed further with the pre-relativistic development of Electromagnetism let us cast the entire structure into a relativistic framework and use Relativity to deduce the standard results of electromagnetic theory. Before proceeding with that, we need to look at the essential structure of the theory as presented by Maxwell, and at his deduction of the existence of electromagnetic waves in the absence of any charge density.

The equations given by Eqs. (6.5), (6.10), (6.17) and (6.18) are known as Maxwell's equations. We consider the case when $\rho$ and $\mathbf{j}$ are zero. Differentiate Eq. (6.17) with respect to time and use Eq. (6.18). Thus

$$\partial^2 \mathbf{E}/\partial t^2 = -(1/\varepsilon_0\mu_0)\nabla \wedge (\nabla \wedge \mathbf{E}) \ . \tag{6.23}$$

Now the right side may be simplified by using the identity

$$\nabla \wedge (\nabla \wedge \mathbf{E}) = \nabla(\nabla \cdot \mathbf{E}) - \nabla^2 \mathbf{E} \ . \tag{6.24}$$

Using Eqs. (6.24) and (6.5) with $\rho = 0$ in Eq. (6.23) we obtain

$$\partial^2 \mathbf{E}/\partial t^2 = (1/\varepsilon_0\mu_0)\, \nabla^2 \mathbf{E} \ . \tag{6.25}$$

This is the wave equation with the speed being the square root of the coefficient of the right side term. The value of the speed turns out to be the speed of light in vacuum.

Differentiating Eq. (6.18) with respect to time and using Eqs. (6.17) and (6.10) with $\mathbf{j} = 0$, and the identity given by Eq. (6.24) for $\mathbf{H}$, we see that

$$\partial^2 \mathbf{H}/\partial t^2 = (1/\varepsilon_0 \mu_0)\, \nabla^2 \mathbf{H} \ . \tag{6.26}$$

Thus $\mathbf{E}$ and $\mathbf{H}$ satisfy the wave equation with the wave speed being the speed of light

$$c = 1/\sqrt{\varepsilon_0 \mu_0} \ . \tag{6.27}$$

It is easily checked that since all materials have higher values of $\varepsilon$ and $\mu$ the speed of the wave in media will be *less* than the speed of light in vacuum. This is true for light as well. It is obvious from here that we should identify light as an electromagnetic wave. This identification has since been verified by producing light by varying electric and magnetic fields appropriately.

## 5. The Four-Vector Formulation

We can write the Maxwell equations in terms of tensors in Minkowski space. The simplest to see is the 4-vector current density which can be written as ($\rho_0$ being the rest charge density)

$$\begin{aligned} j^\mu &= \rho_0 v^\mu = (\gamma \rho_0 c,\ \gamma \rho_0 \mathbf{v}) = (\rho_0' c,\ \rho_0' \mathbf{v}) \\ &= (\rho c, \rho \mathbf{v}) = (j^0, \mathbf{j}) \ , \end{aligned} \tag{6.28}$$

where $\rho$, the moving charge density, increases due to reduction of volume by the $\gamma$-factor. This already gives the moving charge (or charge density). Clearly the current appears as only a charge density in the rest-frame of the charges, i.e., there is no current in a comoving frame. Notice that $j^\mu$ is a vector *density* and not a true vector in that it will not transform as a vector, but there will be an additional, Jacobian factor multiplying the transformation.

Notice that there is a scalar electric potential and a vector magnetic potential. We could try putting them together in a 4-vector electromagnetic potential

$$A_\mu = (\phi/c, -\mathbf{A}) = (A_o, A_i) \tag{6.29}$$

We define the ***Maxwell field tensor*** by

$$F_{\mu\nu} = \partial A_\nu / \partial x^\mu - \partial A_\mu / \partial x^\nu$$
$$\equiv A_{\nu,\mu} - A_{\mu,\nu} \equiv 2A_{[\nu,\mu]} = -F_{\mu\nu} \ . \qquad (6.30)$$

From Eq. (6.30), putting $\mu = i$ and $\nu = j$, we see that

$$F_{ij} = A_{j,i} - A_{i,j} = \varepsilon_{ijk} \, B^k \ , \qquad (6.31)$$

where $\varepsilon_{ijk}$ is the Levi-Civita symbol, which is $+1$ if $i, j, k$ are an even permutation of 1,2,3, is $-1$ if they are an odd permutation and zero if they are not a permutation. Thus, the *spatial* part of the Maxwell field tensor is

$$\mathbf{B} = \mu_0 \mathbf{H} \ . \qquad (6.32)$$

Consider, now, the case where $\mu = 0, \ \nu = i$. Here

$$F_{oi} = \frac{1}{c} \left( \partial A_i / \partial t - \partial \phi / \partial x^i \right) = \frac{1}{c} \, E_i \qquad (6.33)$$

Thus the space-time mixed part of the Maxwell field tensor is the electric field divided by $c$. It should be pointed out that Eq. (6.19) only applies in the absence of a magnetic field. It is clear from Eq. (6.18) that in the absence of a purely electric potential

$$\mathbf{E} = -\partial \mathbf{A} / \partial t \ . \qquad (6.34)$$

We see that the Maxwell tensor contains the electric and magnetic fields,

$$F_{\mu\nu} = \begin{pmatrix} 0 & E_1/c & E_2/c & E_3/c \\ -E_1/c & 0 & B_3 & -B_2 \\ -E_2/c & -B_3 & 0 & B_1 \\ -E_3/c & B_2 & -B_1 & 0 \end{pmatrix} \ . \qquad (6.35)$$

We can raise the indices of the tensor by using the metric tensor

$$F^{\alpha\beta} = g^{\alpha\mu} g^{\beta\nu} \, F_{\mu\nu} \ , \qquad (6.36)$$

where $g^{\alpha\mu}$ is the inverse metric tensor defined by

$$g^{\alpha\mu} g_{\mu\nu} = \delta^{\alpha}_{\nu} \ . \tag{6.37}$$

If we use Cartesian coordinates $g_{\mu\nu}$ is given by Eq. (4.9). In that case

$$g^{00} = 1 = -g^{11} = -g^{22} = -g^{33}, \ \ g^{\alpha\mu} = 0 \text{ if } \alpha \neq \mu \ . \tag{6.38}$$

Thus, we get

$$F^{\alpha\beta} = \begin{pmatrix} 0 & -E_1/c & -E_2/c & -E_3/c \\ E_1/c & 0 & B_3 & -B_2 \\ E_2/c & -B_3 & 0 & B_1 \\ E_3/c & B_2 & -B_1 & 0 \end{pmatrix} . \tag{6.39}$$

The *dual field tensor* is defined by

$$^*F^{\rho\pi} = \frac{1}{2} \varepsilon^{\mu\nu\rho\pi} F_{\mu\nu} \ , \tag{6.40}$$

where $\varepsilon^{\mu\nu\rho\pi}$ is the four dimensional Levi-Civita symbol defined as before for 0, 1, 2, 3. Thus, for example

$$^*F^{01} = \frac{1}{2} \varepsilon^{0123} F_{23} + \frac{1}{2} \varepsilon^{0132} F_{32} = F_{23} \ . \tag{6.41}$$

The complete dual tensor is easily seen to be

$$^*F^{\rho\pi} = \begin{pmatrix} 0 & B_1 & B_2 & B_3 \\ -B_1 & 0 & E_3/c & -E_2/c \\ -B_2 & -E_3/c & 0 & E_1/c \\ -B_3 & E_2/c & -E_1/c & 0 \end{pmatrix} . \tag{6.42}$$

From Eqs. (6.35) and (6.39), we see that

$$\begin{aligned} F^{\mu\nu} F_{\mu\nu} &= -2(\mathbf{E} \cdot \mathbf{E}/c^2 - \mathbf{B} \cdot \mathbf{B}) \\ &= -2\mu_0 (\varepsilon_0 \, \mathbf{E} \cdot \mathbf{E} - \mathbf{B} \cdot \mathbf{B}/\mu_0) \ . \end{aligned} \tag{6.43}$$

Now it can be seen from the theory of Electromagnetism that the free energy density of the electromagnetic field is given by

$$\mathcal{L}_{\text{e.m.}} = \frac{1}{2}(\mathbf{E}\cdot\mathbf{D} - \mathbf{B}\cdot\mathbf{H}) = \frac{1}{2}(\varepsilon_0\,\mathbf{E}\cdot\mathbf{E} - \mathbf{B}\cdot\mathbf{B}/\mu_0) \ . \tag{6.44}$$

Thus, we see that the free energy density for the electromagnetic field, i.e., the electromagnetic Lagrangian is

$$\mathcal{L}_{\text{e.m.}} = -\frac{1}{4}\mu_0\, F^{\mu\nu} F_{\mu\nu} \ . \tag{6.45}$$

From Eqs. (6.35) and (6.42), we obtain

$${}^*F^{\mu\nu} F_{\mu\nu} = 4\mathbf{E}\cdot\mathbf{B}/c \ . \tag{6.46}$$

If the electric and magnetic fields are perpendicular to each other the dual field tensor will be orthogonal to the field tensor.

The force exerted on a test particle of charge $q$ is

$$\mathbf{F} = q\mathbf{E} \ , \tag{6.47}$$

or in terms of the field tensor

$$F_i = qcF_{oi} \ .$$

In general, then, for an arbitrary velocity 4-vector

$$\begin{aligned} F_i &= q\, F_{\mu i}\, v^\mu \\ &= q(F_{oi}\,\gamma c + F_{ji}\,\gamma v^j) \\ &= q'(\mathbf{E} + \mathbf{v}\wedge\mathbf{B}) \ . \end{aligned} \tag{6.48}$$

This is known as the Lorentz force law. It can be seen to be a kinematic effect. It further clarifies the sense in which a magnetic field is just a 'moving electric field'. The magnetic effect of moving charges is related to the electric effect of stationary charges in the same way.

The observable quantity is the field tensor and not the potential giving rise to it. The potential giving a field tensor is not unique. In fact, if we transform the potential by

$$A_\mu \to \widetilde{A}_\mu = A_\mu + f_{,\mu} \ , \tag{6.49}$$

for any scalar function $f$, the field will be transformed by

$$F_{\mu\nu} \to \widetilde{F}_{\mu\nu} = F_{\mu\nu} + 2f_{,[\mu\nu]} = F_{\mu\nu} \ . \tag{6.50}$$

The transformation given by Eq. (6.49) is called a ***gauge transformation***. Equation (6.50) expresses the fact that the field is ***gauge invariant***, i.e. invariant under gauge transformations. This fact is intimately connected with charge conservation. It also plays a pivotal role in the Glashow-Salam-Weinberg theory of electroweak unification and the various attempts at extending the unification to include strong interactions as well.

## 6. The Maxwell Equations Again

It can be checked that the Maxwell equations can be written in the 4-vector formalism as

$$F^{\mu\nu}{}_{,\nu} = \mu_0 j^\mu \ , \tag{6.51}$$

$$F_{[\mu\nu,\rho]} = 0 \ . \tag{6.52}$$

We start with $\mu = 0$ in Eq. (6.51). This clearly yields Eq. (6.5). Now consider $\mu = i$, $\nu = j$ and $\rho = k$ in Eq. (6.52). This immediately gives Eq. (6.10), since we get

$$\frac{1}{3}\left(F_{ij,k} + F_{jk,i} + F_{ki,j}\right) = 0 \ , \quad (i \neq j \neq k \neq i) \ . \tag{6.53}$$

Now, since $k \neq i, j$ and $F_{ij}$ is $B^k$ with $k \neq i, j$, Eq. (6.53) is just Eq. (6.10).

Let us now consider $\mu = i$ in Eq. (6.51). Thus

$$\begin{aligned} F^{io}{}_{,o} + F^{ij}{}_{,j} &= j^i \\ &= -\frac{1}{c}\frac{\partial E^i}{\partial t} + \varepsilon^{ijk} B_{k,j} \ . \end{aligned} \tag{6.54}$$

Thus we get, by dividing through by $\mu_0$,

$$\frac{1}{\mu_0}\nabla \wedge \mathbf{B} = \varepsilon_0\, \partial\mathbf{E}/\partial t + \mathbf{j}\ . \tag{6.55}$$

This is Eq. (6.17). Further, take $\mu = i$, $\nu = j$, $\rho = 0$ in Eq. (6.52). Thus

$$\begin{aligned} F_{ij,o} + F_{jo,i} + F_{oi,j} &= 0 \\ &= \frac{1}{c}\left(\frac{\partial}{\partial t}\,\varepsilon_{ijk}\, B^k + E_{i,j} - E_{j,i}\right)\ . \end{aligned} \tag{6.56}$$

This gives Eq. (6.18).

Consider, now, the divergence of Eq. (6.51)

$$F^{\mu\nu}{}_{,\mu\nu} = -\mu_0 j^\mu{}_{,\mu}\ . \tag{6.57}$$

The left side is zero since we have

$$F^{\mu\nu}{}_{,\mu\nu} = F^{\mu\nu}{}_{,\nu\mu} = -F^{\nu\mu}{}_{,\nu\mu}\ , \tag{6.58}$$

but the dummy indices $\nu$, $\mu$ can be interchanged without making any difference. Hence

$$F^{\mu\nu}{}_{,\mu\nu} = -F^{\mu\nu}{}_{,\mu\nu}\ , \tag{6.59}$$

which implies that

$$F^{\mu\nu}{}_{,\mu\nu} = 0\ . \tag{6.60}$$

Thus, we have, from Eq. (6.57)

$$0 = j^\mu{}_{,\mu} = j^0{}_{,0} + j^i{}_{,i} = \frac{\partial\rho}{\partial t} + \nabla\cdot\mathbf{j}\ , \tag{6.61}$$

i.e., the equation of continuity follows from the Maxwell equations.

We see that electromagnetic theory is naturally expressable in terms of the 4-vector formalism and is consistent with Special Relativity. Electricity and magnetism are different facets of the same fundamental force, which transform into each other due to relative motion. It is found that the inter-relationships are not only more elegantly expressed here, but are also more transparent. The effect of electric and magnetic fields on point charges is easily derivable by changing from the rest-frame to a moving frame. In effect, the electric field in the moving frame is given by the Lorentz transformation

$$E'_i = \gamma(E_i + \varepsilon_{ijk}\, v^j B^k) \ . \tag{6.62}$$

We will now deal with *dynamical* effects in Electromagnetism. To be able to do so we need to discuss acceleration in Special Relativity. We shall proceed to do so in the next chapter.

## Exercise 6

1. Write the electromagnetic 4-vector potential for a point charge and use the Lorentz transformations to work out the 4-vector and hence the electromagnetic field potential for a charge moving with a velocity $\mathbf{v} = (c/2, c/3, c/4)$.

2. Given a constant electric field, $E$, in the $x$-direction and a constant magnetic field, $B$, in the $y$-direction, work out the acceleration of a particle of charge $q$ moving at some instant with a velocity $\mathbf{v} = (c/3, c/5, c/4)$, if the mass of the particle is $m$.

# Chapter 7

# SPECIAL RELATIVITY WITH SMALL ACCELERATIONS

We have seen how to deal with kinematic problems in a consistent way. This treatment led to the development of the Special Theory of Relativity. However, dynamics cannot be dealt with really consistently by Special Relativity. Why not? Because in that case we must deal with accelerations, while Special Relativity is geared only to unaccelerated frames. Of course, we could deal with what accelerations look like to inertial observers, but we will not be able to go into the accelerated frame. We shall see the problems that can arise by including accelerated frames. Nevertheless, it should be possible to deal with the effects of acceleration as 'corrections' of the Special relativistic effects, provided the acceleration is 'sufficiently small'. We will, of course, need to define what is meant by 'sufficiently small accelerations' very precisely. We will then obtain some results for relativistic dynamics. It should be pointed out that for a consistent treatment of dynamics we will need the General Theory of Relativity which deals with arbitrary motion. This theory is presented more fully in a separate volume.

## 1. Some 'Paradoxes' in Special Relativity

A very famous 'paradox' was constructed to try to prove that relativistic time-dilation could not be true. It was variously known as the 'twin paradox' or the 'clock paradox'. The only thing wrong with it (as we shall see) is that there is no paradox according to

Special Relativity, but merely a result which would not have been expected on the basis of everyday experience.

The supposed paradox conceives of a pair of twins, one of whom, $A$, stays at home on the Earth and the other, $B$, visits a distant star at near light speed. Now, according to $A$ the clocks of $B$ run slow and so, when $B$ returns, he will have aged much less than $A$. The point of the 'paradox' is that $B$ could have regarded himself as being at rest and $A$ as moving. Thus, the argument goes, $B$ should expect $A$ to have aged much less. We could replace the twins by 'twin atomic clocks' and the space-trip to a distant star by many trips around the Earth in a fast jet. This experiment has been performed and $B$ *does* run slower than $A$. Since the experiment has been performed and it has given a positive result no paradox is possible. Let us now look at the error in the argument.

The error may be seen by analogy with two people $A$ and $B$ who go from point $P$ to $Q$. $B$ goes via $R$ while $A$ goes directly (see Fig. 21) through point $S$, corresponding to $R$. $A$, of course, traverses a shorter distance than $B$. Now, suppose that we are dealing with a space-time diagram. The 'path of $A$' is simply the passage of time at the same place in his frame. According to him $B$ moves at a speed $v < c$. He moves in one direction up to $R$ and then in the opposite direction from $R$ to $Q$ at the same speed. Now let us go back to the purely spatial picture as seen by $B$. As he goes to $R$ he sees $A$ staying behind him. When he reaches $R$ he sees $A$ at $T_1$. Now he changes direction instantaneously. Suddenly he sees $A$ ahead of him at $T_2$. Of course, $A$ has not moved in the instant and is still at point $T$. It is just $B$'s description of $T$ which has changed. This argument applies equally for the space-time diagram. As $B$ goes away he sees the clocks of $A$ running slow. With his instantaneous acceleration the clocks of $A$ suddenly jump ahead of $B$'s clocks and end up ahead of his. Therefore, there is no paradox.

Another 'paradox' was invented to demonstrate the errors arising from dealing with forces in Special Relativity and the methods that should be used to avoid such errors. In this 'paradox' (see Fig. 22a) an observer on a smooth table-top which has a hole of diameter $\ell$ in

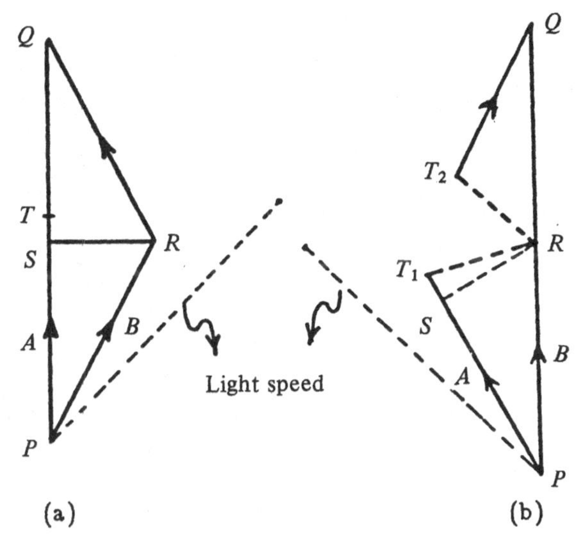


Fig. 21. The resolution of the 'clock paradox'. (a) Imagine $A$ going from $P$ to $Q$ via $S$ while $B$ goes via $R$ along a straight line. (b) According to $B$, if he insists that he is proceeding along a straight line, when he turns around at $R$ he sees $A$ suddenly jumping from the point $T_1$ behind him to the point $T_2$ ahead of him. Actually there is only one point $T$ seen in two ways by $B$ as he changes his definition of 'in-front', i.e., his direction. If we take these as space-time diagrams we have the resolution of the clock paradox. Clearly $PT + TQ \neq PR + PQ$ and $PT_1 + T_2Q \neq PR + RQ$.

it sees a thin rod of rest-length $\ell$ coming towards the hole at speed $v$. He reasons that since the rod has length $\ell' = \ell/\gamma < \ell$, it will fall through the hole. An observer comoving with the rod argues that the hole has a diameter $\ell'$ while the rod is of length $\ell > \ell'$. Thus, the rod is too long to go through the hole and it will pass over it. To decide which has happened a paper wall is put up along the line of motion of the rod on the table-top beyond the hole. If the table-top observer, $O$, is correct, the wall will remain intact. If the observer who is comoving with the rod, $O'$, is correct, the wall will be broken. Thus, we can test which one is correct. There can be no paradox in this case either. The only question is 'Which one is wrong?'

The answer must obviously be that $O'$ is wrong since his frame changes when he comes over the hole. $O$ continues in a non-

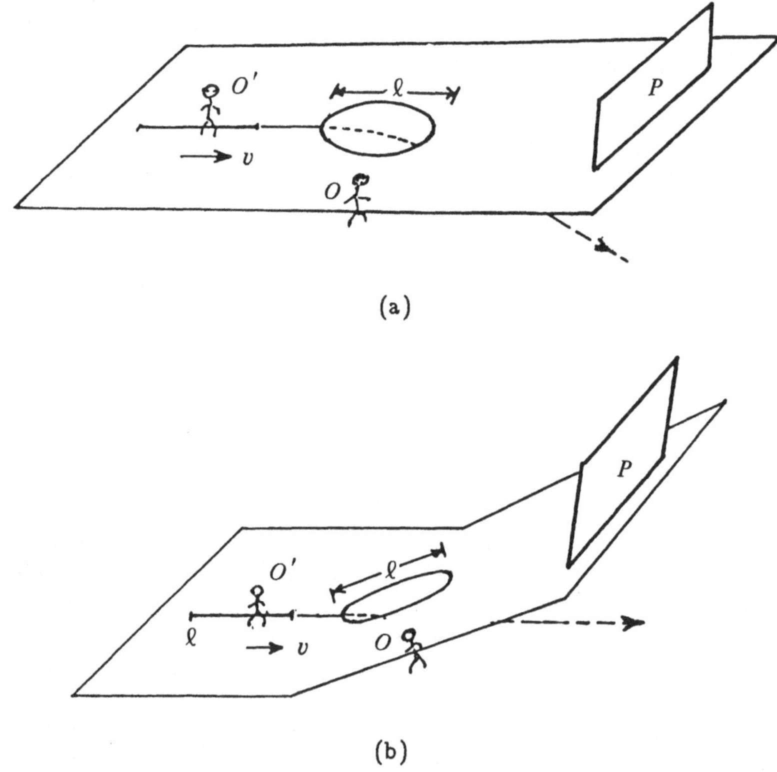


Fig. 22. (a) An observer $O$ sees the rod of rest-length $\ell$ coming towards a hole of diameter $\ell$ and predicts that the rod will fall through the hole due to the Lorentz contraction due to the speed $v$. The paper wall, $P$, stays intact. (b) The observer $O'$, riding on the rod does not agree till he comes to the hole when his frame changes and he sees the table-top curving up in front of him and he is able to go through the hole, leaving $P$ intact, even though the diameter appears to him to be less than $\ell$.

inertial frame. $O'$ is in a non-inertial frame initially, in that he can see objects fall when released. However, over the hole, as he falls, he sees objects float in his frame, since they are falling at the same rate as he is. If he insists that he is moving in a straight line as he continues, he must observe the table-top curve up in front of him (Fig. 22b) and go through the hole that has risen up in front of him. Thus there is, again, no paradox. Again, geometrical arguments are used to resolve the problem. Clearly, the consistent relativistic treatment of acceleration must be geometrical.

## 2. The Range of Validity of Special Relativity

One may raise the question whether Special Relativity has to be discarded if we deal with accelerated frames. Even on the Earth we do not have a real non-accelerated, i.e., inertial, frame. In fact it is a moot point whether a genuinely inertial frame can exist in the Universe. The gravitational field of the object itself will cause some non-inertial effects. However, it is clear that for any given accuracy of observation there will be some sufficiently small acceleration which could be considered negligible. The question is "How small must the acceleration be?"

For Special Relativity without acceleration we know that relativistic effects only come in at speeds comparable to the speed of light. Thus, Classical Mechanics holds to the extent that $v^2 \ll c^2$ and we can use relativistic corrections as $O(v/c)^2$ modifications of the classical result. Now, given one inertial frame and another non-inertial frame we can get an estimate of the speed attained by the non-inertial frame relative to the inertial frame if they started at rest initially by taking the acceleration, a, to be constant,

$$\frac{1}{2}v^2 \approx \mathbf{a}\cdot\mathbf{x} \ , \tag{7.1}$$

where $\mathbf{x}$ is the displacement vector through which the acceleration has been applied. Thus the acceleration is negligible if

$$\mathbf{a}\cdot\mathbf{x} \ll c^2/2 \ . \tag{7.2}$$

Clearly, we can obtain 'corrections' of the special relativistic results due to acceleration as $O(\alpha)$ terms, where

$$\alpha = 2\mathbf{a}\cdot\mathbf{x}/c^2 \ll 1 \ . \tag{7.3}$$

It is to be noticed that for the 'clock paradox' we cannot make $\alpha \ll 1$. This is so because the spaceship travels at high speeds (i.e., $v^2/c^2$ must not be negligible) and the acceleration is enough to change the velocity from $v$ to $-v$. If we make $|\mathbf{x}|$ small we will have to make a correspondingly large but $\mathbf{a} \cdot \mathbf{x}$ will remain of the order of $v^2$. Thus the only way to make $\alpha$ negligible is to make $v^2/c^2$ negligible. In this case the relativistic effect is anyhow undetectable.

In the 'paradox' of the table with a hole, we see that if there were no acceleration due to gravitation, the rod would not fall anyhow. If there is some acceleration it would still not fall through if there were a finite thickness of the rod. Thus, the acceleration could only be negligible if the distance fallen was much less than the thickness of the rod. For an ideal, infinitely thin rod there could not be any sufficiently small acceleration. Thus, the result would remain true anyhow (that the rod will fall through the hole).

## 3. The Gravitational Red-Shift

Consider a photon of frequency $\nu$ rising a distance $d$ from a gravitating source of mass $M$, having started at a distance $r$. The photon has an effective mass

$$m = E/c^2 = h\nu/c^2 \ . \tag{7.4}$$

Thus, the work done in moving the photon up is

$$W = \int_r^{r+d} \frac{GMm}{\bar{r}^2}\, d\bar{r} = \frac{h\nu GM}{c^2}\left(\frac{1}{r} - \frac{1}{r+d}\right) \ . \tag{7.5}$$

This energy can come only from the photon itself. Thus if the energy loss is small we can get

$$h\delta\nu \approx \frac{h\nu GM}{c^2}\,\frac{d}{r(r+d)} \ . \tag{7.6}$$

Thus the decrease in frequency per unit frequency, if $d \ll r$, is

$$\frac{\delta\nu}{\nu} \approx \frac{GM}{c^2 r}\,\frac{d}{r} \ . \tag{7.7}$$

For the argument to be valid we have

$$\alpha = GM/c^2 r \ll 1 \ . \tag{7.8}$$

The gravitational red-shift predicted here has been measured by sending light up a 70 metre tower on the surface of the Earth. It is found to be exactly the predicted value within the limits of experimental error.

## 4. Gravitational Deflection of Light

Newton, a long time ago, had suggested that light should be subject to his gravitational law just as matter was. However, his suggestion had not been taken very seriously because light was generally regarded as a wave – very different from matter. Einstein, who had revived Newton's corpuscular theory in a different form, was perfectly ready to try to work out the gravitational deflection of the path of light. The calculation was later repeated using General Relativity and gave a value twice the value obtained here. The full relativistic result, i.e. that obtained later, was found to be correct within experimental errors. We shall discuss the experiment which tested the prediction later. First we consider an idealised 'thought experiment'.

Consider an experimenter in a lift as shown in Fig. 23a. A beam of light starts from $A$ and is received at $B$. The width of the lift is $s$. Thus the light takes a time $t = s/c$ to cross the lift. Now, if the experimenter releases a ball in the air it falls (demonstrating that he is in a non-inertial frame). Having performed this experiment he has a friend release the lift so that it falls freely (see Fig. 23b). Now when he drops the ball it floats (demonstrating that he is in an inertial frame) since it falls at the same rate as he does. In the time $t$ it falls a distance

$$d = \frac{1}{2} g t^2 = g s^2/2c^2 \ . \tag{7.9}$$

Now, the results are valid if

$$\alpha = g s/2c^2 \ll 1 \ . \tag{7.10}$$

Thus we have deflection of the light path by an angle

$$\theta \approx \tan\theta = d/s = (g s/2c^2) \ . \tag{7.11}$$

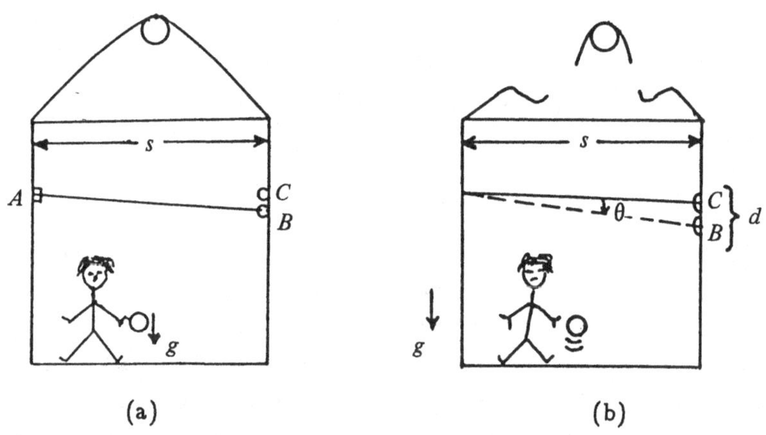


Fig. 23. An experimenter in a freely falling lift. (a) At first the lift is fixed, light goes across the lift (a width $s$) from $A$ to $B$. Also a ball released by the experimenter drops. (b) Now the lift is allowed to fall freely. In the time the light crosses the lift, it falls a distance $d$ so that the light falls at point $C$ instead of $B$. Also the ball released falls with the lift and hence appears to the experimenter to float. Thus (b) gives an inertial frame while (a) gives an accelerated frame.

Of course, it would not be possible to test this result by a terrestrial experiment as, even if $s = 100$ metre, $\theta \approx 0.5 \times 10^{-14}$ rad $\approx$ $10^{-8}$ arc seconds. However, generally we have

$$g = GM/R^2 \ , \tag{7.12}$$

which gives, for any gravitating source of mass $M$ at distance $R$,

$$\theta \approx \frac{GMs}{2c^2 R^2} \ . \tag{7.13}$$

Now consider light grazing by the Sun. In that case $R$ is the radius of the Sun and $s$ is approximately the solar diameter, $2R$. Thus, for the Sun

$$\theta \approx \frac{GM}{c^2 R} \ . \tag{7.14}$$

The problem now is how to see light grazing by the Sun? The light source would have to be a star. It is not normally possible to see a star at the edge of the Sun. However, due to the coincidence that

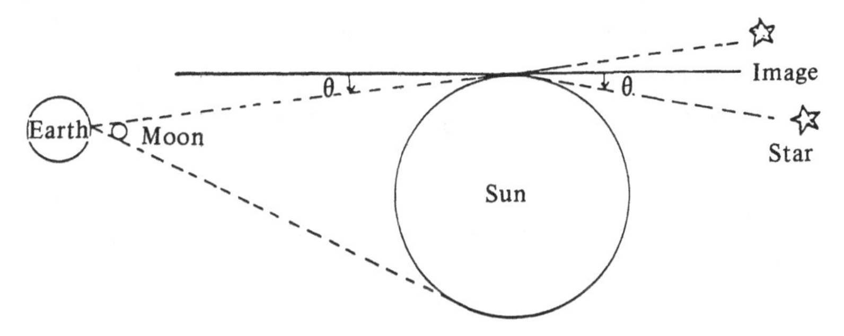


Fig. 24. A star behind the disc of the Sun will appear to be just away from it due to the gravitational deflection of light. Here, of course, nothing is drawn to scale. The point is simply that the Sun is just blotted out by the Moon at the point of observation, while the star that was blotted out has become visible. Notice that the effective deflection is $2\theta$ and not $\theta$.

the lunar and solar angular diameters are equal (both $\frac{1}{2}^\circ$) during a total solar eclipse, the disc of the Sun is just blotted out. At that time light that grazes by the Sun can just be seen. Now, as shown in Fig. 24, light from a star that would be behind the disc of the Sun will appear to come from a different point. By comparing the position of the star with respect to other stars (see Fig. 25), Eddington was able to measure the gravitational deflection of light. The result obtained was four times that given by Eq. (7.14). As is clear from Fig. 24, the quantity measured is $2\theta$ and not $\theta$. Thus, Eddington found a value twice that given by the above argument. As mentioned earlier, the correct value is predicted by General Relativity.

## 5. Four-Vector Acceleration and Force

If we can totally ignore the problem of having non-inertial frames when dealing with accelerations, i.e., if we are considering what the acceleration seen by one unaccelerated observer appears to another accelerated observer, we can continue with the use of the four-vector formalism. We can, then, define the *4-vector acceleration*

$$\mathcal{A}^\mu = \frac{dv^\mu}{d\tau} = \frac{d}{d\tau}\left(\frac{c}{\sqrt{1-\mathbf{v}\cdot\mathbf{v}/c^2}}\,,\ \frac{\mathbf{v}}{\sqrt{1-\mathbf{v}\cdot\mathbf{v}/c^2}}\right)\ . \tag{7.15}$$

Now, if we define the usual acceleration vector by

$$\mathbf{a} = d\mathbf{v}/dt = (d\mathbf{v}/d\tau)(d\tau/dt)\ , \tag{7.16}$$

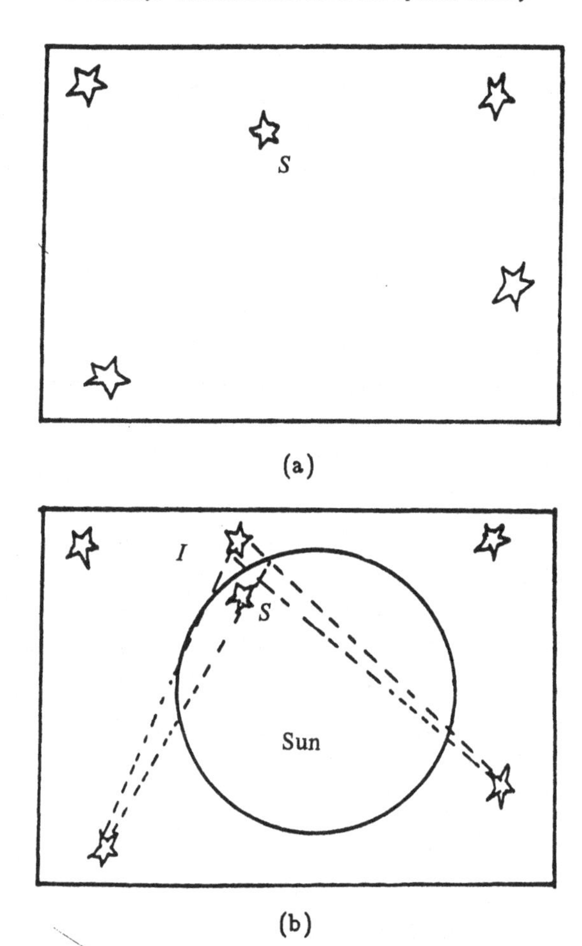


Fig. 25. Eddington's observation of the gravitational deflection of light (not drawn to scale). (a) The positions of some stars relative to each other. In particular the position of $S$ relative to the other stars is seen. (b) Later, during a solar eclipse, when $S$ should have been hidden at the edge of the solar disc, it appears to be seen at $I$. The position of $I$ is measurably different from $S$. The deflection was twice the value predicted by Special Relativity and exactly the value expected on the geometrical arguments of General Relativity.

we see that

$$d\mathbf{v}/d\tau = \mathbf{a}\, dt/d\tau = \frac{\mathbf{a}}{\sqrt{1-\mathbf{v}\cdot\mathbf{v}/c^2}} \ . \tag{7.17}$$

Thus we have

$$\mathcal{A}^\mu = \left(\frac{1}{c}\frac{\mathbf{v}\cdot d\mathbf{v}/d\tau}{(1-\mathbf{v}\cdot\mathbf{v}/c^2)^{3/2}}\,,\ \frac{d\mathbf{v}/d\tau}{\sqrt{1-\mathbf{v}\cdot\mathbf{v}/c^2}}+\frac{\mathbf{v}}{c^2}\frac{\mathbf{v}\cdot d\mathbf{v}/d\tau}{(1-\mathbf{v}\cdot\mathbf{v}/c^2)^{3/2}}\right)$$

$$= \frac{c}{(1-\mathbf{v}\cdot\mathbf{v}/c^2)}\left(\frac{\mathbf{a}\cdot\mathbf{v}/c^2}{(1-\mathbf{v}\cdot\mathbf{v}/c^2)}\,,\ \frac{\mathbf{a}}{c}+\frac{\mathbf{v}}{c}\frac{\mathbf{a}\cdot\mathbf{v}/c^2}{(1-\mathbf{v}\cdot\mathbf{v}/c^2)}\right)\,. \tag{7.18}$$

Notice that in the limit $c \to \infty$

$$\mathcal{A}^\mu \to (0, \mathbf{a})\ , \tag{7.19}$$

i.e. we have the usual acceleration as the classical limit of the acceleration 4-vector. The factor $(1-\mathbf{v}\cdot\mathbf{v}/c^2)^{-1}$ comes from the time dilation which is squared, since acceleration is the *second* time derivative. Let us call this classical limit of the 4-vector acceleration $\mathcal{A}^\mu_c$. Then, if we define the relativistic correction of this 4-vector by

$$\mathcal{A}^\mu_R = \mathcal{A}^\mu - \mathcal{A}^\mu_c\ , \tag{7.20}$$

we find that

$$\mathcal{A}^\mu_R = \frac{\mathbf{a}\cdot\mathbf{v}/c^2}{(1-\mathbf{v}\cdot\mathbf{v}/c^2)^2}(c,\mathbf{v}) = \frac{\mathbf{a}\cdot\mathbf{v}/c^2}{(1-\mathbf{v}\cdot\mathbf{v}/c^2)^{3/2}}\,v^\mu\ . \tag{7.21}$$

Another way of expressing the change is by writing $\mathcal{A}^\mu$ as

$$\mathcal{A}^\mu = \frac{1}{(1-\mathbf{v}\cdot\mathbf{v}/c^2)^2}\left(\mathbf{a}\cdot\mathbf{v}/c, \mathbf{a}+(\mathbf{a}\wedge\mathbf{v})\wedge\mathbf{v}/c^2\right)\ . \tag{7.22}$$

Thus we see that the classical dynamic result is altered by the square of the factor $(1-\mathbf{v}\cdot\mathbf{v}/c^2)$ and by the addition of a vector triple product. Clearly, if $\mathbf{a}\perp\mathbf{v}$ there is no 'correction' in the zero-component, while if $\mathbf{a}\|\mathbf{v}$ the extra term in the space-component disappears.

The *4-vector force* is then defined by

$$\mathcal{F}^\mu = m\mathcal{A}^\mu = \left(\frac{m'\mathbf{a}'\cdot\mathbf{v}/c}{1-\mathbf{v}\cdot\mathbf{v}/c^2}\,,\ m'\mathbf{a}'+\frac{\mathbf{v}}{c}\frac{m'\mathbf{a}'\cdot\mathbf{v}/c}{1-\mathbf{v}\cdot\mathbf{v}/c^2}\right)$$

$$= \mathcal{F}^\mu{}_c + \frac{m'\mathbf{a}'\cdot\mathbf{v}/c^2}{\sqrt{1-\mathbf{v}\cdot\mathbf{v}/c^2}}\,v^\mu\ , \tag{7.23}$$

where $\mathbf{a}'$ is the relativistically corrected acceleration, $d\mathbf{v}/d\tau$, and

$$\mathcal{F}^{\mu}{}_{c} = (O, \mathcal{F}_c) = (O, m'\mathbf{a}') \ . \tag{7.24}$$

Thus the relativistic correction is

$$\mathcal{F}^{\mu}{}_{R} = \frac{\mathcal{F}_c \cdot \mathbf{v}/c^2}{\sqrt{1 - \mathbf{v}\cdot\mathbf{v}/c^2}}\, v^{\mu} \ . \tag{7.25}$$

The question arises as to the interpretation of the zero component of the force 4-vector. Now,

$$\begin{aligned} \mathcal{F}^{\mu} &= dp^{\mu}/d\tau = (d/d\tau)(E/c, \mathbf{p}) \\ &= \left(\frac{1}{c}\frac{dE}{d\tau}, \frac{d\mathbf{p}}{d\tau}\right) \ . \end{aligned} \tag{7.26}$$

Thus, we see that the zero component corresponds to a rate of change of energy of the particle. Hence, we see that, in general, an accelerated particle either radiates or absorbs energy according as $dE/d\tau \gtrless 0$. We have

$$\dot{E} = \frac{\mathcal{F}_c \cdot \mathbf{v}}{(1 - \mathbf{v}\cdot\mathbf{v}/c^2)} \ . \tag{7.27}$$

Clearly, if $\mathcal{F}_c \perp \mathbf{v}$, as is the case in circular motion, no such radiation is to be expected. It is necessary to point out, here, that we are *not* dealing with motion of charged matter in the presence of electromagnetic fields, but of neutral matter under mechanical forces only.

To deal with accelerated charges we extend Eq. (6.47) to the 4-vector force,

$$\mathcal{F}^{\mu} = qF^{\mu}{}_{\nu}\, v^{\nu} \ . \tag{7.28}$$

Thus, we have the zero component giving

$$\dot{E} = qc\, F^{o}{}_{i}\, v^{i} = q'c\,\mathbf{E}\cdot\mathbf{v} \ . \tag{7.29}$$

Now, also

$$\begin{aligned} dE/dt &= (dE/d\tau)(d\tau/dt) = \gamma^{-1}\, dE/d\tau \\ &= \gamma^{-1}\,\dot{E} \ . \end{aligned} \tag{7.30}$$

Thus we get

$$dE/dt = q'\mathbf{E}\cdot\mathbf{v}/c \ . \tag{7.31}$$

If a particle of charge $q$ moving at speed $v$ is opposed by an electric field it will radiate energy and lose speed. If, on the other hand, the electric field is in the direction of motion, it will absorb energy from the field and accelerate. This is the principle on the basis of which charged particles are accelerated.

## 6. Restatement of the Principle of Special Relativity

In view of the fact that we have to deal with non-inertial frames as well, we need to restate the principle of special relativity in a consistent way. This may be managed as follows: "Relatively unaccelerated frames are physically equivalent". This formulation is supported by the fact that the effect of acceleration on each frame will be the same to the lowest order. Thus all physical laws will remain invariant. The correction is, of course, velocity dependent, so there will be relativistic corrections which will be different in different, relatively unaccelerated frames.

## 7. Change of Metric Due to Acceleration

It had been shown in Chapter 2, Sec. 5, that the geometry of a frame in uniform circular motion is non-Euclidean. We will now look at the effect geometrically and more generally. (This metric was derived by F. Hussain and A. Qadir in ZAMP *37* (1986) 387.) It is already apparent from Eq. (7.18) that in the case that $\mathbf{a}\perp\mathbf{v}$ there is only a time dilation effect. This fact is also clear from Eqs. (7.23) and (7.24). Here, by writing Eq. (2.57) in differential form, we have

$$\left.\begin{aligned} dt' &= \frac{dt(1-r^2\Omega^2/c^2)+(r\Omega^2/c^2)t\,dr}{(1-r^2\Omega^2/c^2)^{3/2}}\ , \quad dr' = dr\ , \\ d\theta' &= d\theta - \frac{\Omega dt(1-r^2\Omega^2/c^2)+(r\Omega^3 t/c^2)dr}{(1-r^2\Omega^2/c^2)^{3/2}}\ , \quad dz' = dz\ . \end{aligned}\right\} \tag{7.32}$$

Thus the metric becomes

$$\left.\begin{aligned} ds^2 &= c^2\,dt'^2 - dr'^2 - r'^2 d\theta'^2 - dz'^2 \\ &= c^2\,\frac{[dt(1-r^2\Omega^2/c^2)+dr(r\Omega^2 t/c^2)]^2}{(1-r^2\Omega^2/c^2)^3} - dr^2 \\ &\quad - r^2\left(d\theta - \frac{\Omega dt(1-r^2\Omega^2/c^2)+(r\Omega^3 t/c^2)dr}{(1-r^2\Omega^2/c^2)^{3/2}}\right)^2 - dz^2\ . \end{aligned}\right\} \tag{7.33}$$

Writing

$$(1-r^2\Omega^2/c^2)^{-\frac{1}{2}} = \gamma\ , \tag{7.34}$$

we get

$$\begin{aligned} ds^2 &= c^2 dt^2 + 2r\Omega^2 t\gamma^2 dt\,dr + 2r^2\Omega\gamma^4\,dt\,d\theta - (1 - r^2\Omega^4\,\frac{t^2}{c^2}\,\gamma^4)dr^2 \\ &\quad + 2r^3\Omega^3\,\frac{t}{c^2}\,\gamma^6\,dr\,d\theta - r^2\,d\theta^2 - dz^2\ , \end{aligned} \tag{7.35}$$

which gives the metric tensor

$$g_{\mu\nu} = \begin{pmatrix} c^2 & r\Omega^2 t\gamma^2 & r^2\Omega\gamma^4 & 0 \\ r\Omega^2 t\gamma^2 & -(1-r^2\Omega^4\,\frac{t^2}{c^2}\,\gamma^4) & r^3\Omega^3\,\frac{t}{c^2}\,\gamma^6 & 0 \\ r^2\Omega\gamma^4 & r^3\Omega^3\,\frac{t}{c^2}\,\gamma^6 & -r^2 & 0 \\ 0 & 0 & 0 & -1 \end{pmatrix}\ . \tag{7.36}$$

A physically more interesting case is the acceleration of a particle due to gravity. The frame of the accelerated particle is inertial in that objects released in that frame will float. We can regard the frame of a particle that remains fixed, then, as an accelerated frame, while the other frame is unaccelerated. At any instant, the inertial frame will be moving at a speed, say $v$, relative to the fixed (accelerated) frame. Let the coordinates in the moving (inertial) frame be $t'$, $r'$, $\theta'$, $\phi'$ and let the motion be in the radial direction. Then

$$ds^2 = c^2 dt'^2 - dr'^2 - r'^2 d\theta'^2 - r'^2\,\sin^2\theta'\,d\phi'^2\ . \tag{7.37}$$

At any instant there will be length contraction and time dilation between the two frames

$$dr = dr'\,\sqrt{1-v^2/c^2}\ ,\quad dt = dt'/\sqrt{1-v^2/c^2}\ , \tag{7.38}$$

but $r'$ will be unchanged, i.e., $r = r'$, and $\theta = \theta'$, $\phi = \phi'$. Thus

$$ds^2 = c^2(1 - v^2/c^2)dt^2 - \frac{dr^2}{(1 - v^2/c^2)} - r^2(d\theta^2 + \sin^2\theta\, d\phi^2) \ . \quad (7.39)$$

Now, if the fixed frame is taken as being at an infinite distance and the inertial frame as dropping freely from that infinite distance, the kinetic energy will equal the potential energy. If the particle had mass $m$ and was falling in the gravitational field of a particle of mass $M$,

$$\frac{1}{2} mv^2 = GMm/r \ . \quad (7.40)$$

Thus the factor

$$1 - v^2/c^2 = 1 - 2GM/c^2 r \ . \quad (7.41)$$

Hence the metric becomes

$$ds^2 = c^2(1 - 2GM/c^2 r)dt^2 - \frac{dr^2}{(1 - 2GM/c^2 r)} - r^2(d\theta^2 + \sin^2\theta\, d\phi^2) \ . \quad (7.42)$$

This should represent the metric in a gravitational field as effective for a test particle. It is found that the fully consistent treatment of Relativity yields the same result. This metric is known as the *Schwarzschild metric* and the corresponding metric tensor is

$$g_{\mu\nu} = \begin{pmatrix} (1 - 2GM/c^2 r) & 0 & 0 & 0 \\ 0 & -(1 - 2GM/c^2 r)^{-1} & 0 & 0 \\ 0 & 0 & -r^2 & 0 \\ 0 & 0 & 0 & -r^2 \sin^2\theta \end{pmatrix} , \quad (7.43)$$

We see that accelerated motion will be described in terms of non-Euclidean geometries, in the sense that curvatures will be involved. Such geometries are called Riemannian. Of course, the geometry of Special Relativity is anyhow non-Euclidean in the sense of having a different signature, i.e. a Minkowski geometry. The geometry for General Relativity is generally known as pseudo-Riemannian, implying both curvature and the changed signature. A better name, after the Russian who invented such geometries, is 'Lobachevskian geometry'.

We will not, here, discuss the development of the Relativistic theory of accelerated motion, i.e. General Relativity. Nor will we consider, here, the implications of the change of metric given by Eqs. (7.42) and (7.43). However, the reader is encouraged to ponder how to deduce consequences of these results. They will be discussed in more detail in a separate volume.

## Exercise 7

1. Accelerated charged particles emit electromagnetic radiation. What type of radiation will come from electrically neutral matter when it is accelerated?

2. What will be the relativistic correction to a classical force in the $x$-direction at an instant when the velocity of the object is $\mathbf{v} = (c/2, 0, 0)$ and the acceleration is $0.005\,c$ per second if its mass is $m$?

3. Given two equal charges, $q$, of mass $m$ each, coming towards each other with speeds $c/2$ and $-c/2$ in the laboratory frame, at some instant, determine the relativistically corrected forces they exert on each other (where the distance between them at the instant under consideration is $r$).

4. Work out the acceleration 4-vector of a particle in uniform circular motion.

5. The frequency of light received from a quasi-stellar object (a *quasar* or QSO) is three times less than the emitted frequency. If its rest mass is taken to be one solar mass ($2 \times 10^{33}$ g) what is its least speed relative to us, given that the light comes from its surface and its density is $10^8$ g/cc?

6. Work out the rate of change of energy of a particle of mass $m$ and charge $q$ due to a radial acceleration of constant magnitude $a$ if the motion is circular. How is this result altered if the acceleration is linear?

# INDEX